Excel for Statistics

Excel for Statistics is a series of textbooks that explain how to use Excel to solve statistics problems in various fields of study. Professors, students, and practitioners will find these books teach how to make Excel work best in their respective field. Applications include any discipline that uses data and can benefit from the power and simplicity of Excel. Books cover all the steps for running statistical analyses in Excel 2016, Excel 2013, Excel 2010 and Excel 2007. The approach also teaches critical statistics skills, making the books particularly applicable for statistics courses taught outside of mathematics or statistics departments.

Series editor: Thomas J. Quirk

The following books are in this series:

T.J. Quirk, *Excel 2016 for Engineering Statistics: A Guide to Solving Practical Problems*, Excel for Statistics. Springer International Publishing Switzerland 2016.

T.J. Quirk, *Excel 2016 for Business Statistics: A Guide to Solving Practical Problems*, Excel for Statistics. Springer International Publishing Switzerland 2016.

T.J. Quirk, M. Quirk, H.F. Horton, *Excel 2016 for Biological and Life Sciences Statistics: A Guide to Solving Practical Problems*, Excel for Statistics. Springer International Publishing 2016.

T.J. Quirk, M. Quirk, H.F. Horton, *Excel 2016 for Physical Sciences Statistics: A Guide to Solving Practical Problems*, Excel for Statistics. Springer International Publishing 2016.

T.J. Quirk, E. Rhiney, *Excel 2016 for Marketing Statistics: A Guide to Solving Practical Problems*, Excel for Statistics. Springer International Publishing Switzerland 2016.

T.J. Quirk, *Excel 2016 for Educational and Psychological Statistics: A Guide to Solving Practical Problems*, Excel for Statistics. Springer International Publishing Switzerland 2016.

T.J. Quirk, *Excel 2016 for Social Science Statistics: A Guide to Solving Practical Problems*, Excel for Statistics. Springer International Publishing Switzerland 2016.

T.J. Quirk, S. Cummings, *Excel 2016 for Health Services Management Statistics: A Guide to Solving Practical Problems*. Excel for Statistics. Springer International Publishing Switzerland 2016.

T.J. Quirk, J. Palmer-Schuyler, *Excel 2016 for Human Resource Management Statistics: A Guide to Solving Practical Problems*, Excel for Statistics. Springer International Publishing Switzerland 2016.

T.J. Quirk, M. Quirk, H.F. Horton, *Excel 2016 for Environmental Sciences Statistics: A Guide to Solving Practical Problems*, Excel for Statistics. Springer International Publishing Switzerland 2016.

T.J. Quirk, J. Palmer-Schuyler, *Excel 2013 for Human Resource Management Statistics: A Guide to Solving Practical Problems*, Excel for Statistics. Springer International Publishing Switzerland 2016.

T.J. Quirk, S. Cummings, *Excel 2013 for Health Services Management Statistics: A Guide to Solving Practical Problems*. Excel for Statistics. Springer International Publishing Switzerland 2016.

T.J. Quirk, M. Quirk, H.F. Horton, *Excel 2013 for Physical Sciences Statistics: A Guide to Solving Practical Problems*. Excel for Statistics. Springer International Publishing Switzerland 2016.

T.J. Quirk. *Excel 2013 for Educational and Psychological Statistics: A Guide to Solving Practical Problems*, Excel for Statistics. Springer International Publishing Switzerland 2015.

T.J. Quirk, M. Quirk, H.F. Horton, *Excel 2013 for Biological and Life Sciences Statistics: A Guide to Solving Practical Problems*, Excel for Statistics. Springer International Publishing 2015.

T.J. Quirk, *Excel 2013 for Social Science Statistics: A Guide to Solving Practical Problems*, Excel for Statistics. Springer International Publishing Switzerland 2015.

T.J. Quirk, *Excel 2013 for Business Statistics: A Guide to Solving Practical Problems*, Excel for Statistics. Springer International Publishing Switzerland 2015.

T.J. Quirk. *Excel 2013 for Engineering Statistics: A Guide to Solving Practical Problems*, Excel for Statistics. Springer International Publishing Switzerland 2015.

T.J. Quirk, M. Quirk, H.F. Horton, *Excel 2013 for Environmental Sciences Statistics: A Guide to Solving Practical Problems*, Excel for Statistics. Springer International Publishing Switzerland 2015.

Additional Statistics books by Dr. Tom Quirk that have been published by Springer

T.J. Quirk, *Excel 2010 for Engineering Statistics: A Guide to Solving Practical Problems.* Springer International Publishing Switzerland 2014.

T.J. Quirk, S. Cummings, *Excel 2010 for Health Services Management Statistics: A Guide to Solving Practical Problems.* Springer International Publishing Switzerland 2014.

T.J. Quirk, M. Quirk, H. Horton, *Excel 2010 for Physical Sciences Statistics: A Guide to Solving Practical Problems.* Springer International Publishing Switzerland 2013.

T.J. Quirk, M. Quirk, H.F. Horton, *Excel 2010 for Biological and Life Sciences Statistics: A Guide to Solving Practical Problems.* Springer Science+Business Media New York 2013.

T.J. Quirk, M. Quirk, H.F. Horton, *Excel 2007 for Biological and Life Sciences Statistics: A Guide to Solving Practical Problems.* Springer Science+Business Media New York 2013.

T.J. Quirk, *Excel 2010 for Social Science Statistics: A Guide to Solving Practical Problems.* Springer Science+Business Media New York 2012.

T.J. Quirk, *Excel 2010 for Educational and Psychological Statistics: A Guide to Solving Practical Problems.* Springer Science+Business Media New York 2012.

T.J. Quirk, *Excel 2007 for Business Statistics: A Guide to Solving Practical Problems.* Springer Science+Business Media New York 2012.

T.J. Quirk, *Excel 2007 for Social Science Statistics: A Guide to Solving Practical Problems.* Springer Science+Business Media New York 2012.

T.J. Quirk, *Excel 2007 for Educational and Psychological Statistics: A Guide to Solving Practical Problems.* Springer Science+Business Media New York 2012.

T.J. Quirk, *Excel 2010 for Business Statistics: A Guide to Solving Practical Problems.* Springer Science+Business Media 2011.

More information about this series at http://www.springer.com/series/13491

Thomas J. Quirk • Meghan H. Quirk
Howard F. Horton

Excel 2016 for Biological and Life Sciences Statistics

A Guide to Solving Practical Problems

Springer

Thomas J. Quirk
Webster University
St. Louis, MO, USA

Meghan H. Quirk
Bailey, CO, USA

Howard F. Horton
Bailey, CO, USA

Excel for Statistics
ISBN 978-3-319-39488-6 ISBN 978-3-319-39489-3 (eBook)
DOI 10.1007/978-3-319-39489-3

Library of Congress Control Number: 2016941175

Printed on acid-free paper

This Springer imprint is published by Springer Nature
The registered company is Springer International Publishing AG Switzerland

This book is dedicated to the more than 3,000 students I have taught at Webster University's campuses in St. Louis, London, and Vienna; the students at Principia College in Elsah, Illinois; and the students at the Cooperative State University of Baden-Wuerttemberg in Heidenheim, Germany. These students taught me a great deal about the art of teaching. I salute them all, and I thank them for helping me to become a better teacher.

Thomas J. Quirk

We dedicate this book to all the newly inspired students emerging into the ranks of the various fields of science.

Meghan H. Quirk and Howard F. Horton

Preface

Excel 2016 for Biological and Life Sciences Statistics: A Guide to Solving Practical Problems is intended for anyone looking to learn the basics of applying Excel's powerful statistical tools to their science courses or work activities. If understanding statistics isn't your strongest suit, if you are not especially mathematically inclined, or if you are wary of computers, then this is the right book for you.

Here you'll learn how to use key statistical tests using Excel without being overpowered by the underlying statistical theory. This book clearly and methodically shows and explains how to create and use these statistical tests to solve practical problems in the biological and life sciences.

Excel is an easily available computer program for students, instructors, and managers. It is also an effective teaching and learning tool for quantitative analyses in science courses. The powerful numerical computational ability and the graphical functions available in Excel make learning statistics much easier than in years past. However, this is the first book to show Excel's capabilities to more effectively teach science statistics; it also focuses exclusively on this topic in an effort to render the subject matter not only applicable and practical but also easy to comprehend and apply.

Unique features of this book:

- This book is appropriate for use in any course in biological or life sciences statistics (at both undergraduate and graduate levels) as well as for managers who want to improve the usefulness of their Excel skills.
- Includes 164 color screenshots so that you can be sure you are performing the Excel steps correctly.
- You will be told each step of the way, not only *how* to use Excel but also *why* you are doing each step so that you can understand what you are doing and not merely learn how to use statistical tests by rote.
- Includes specific objectives embedded in the text for each concept, so you can know the purpose of the Excel steps.

- This book is a tool that can be used either by itself or along with *any* good statistics book.
- Statistical theory and formulas are explained in clear language without bogging you down in mathematical fine points.
- You will learn both how to write statistical formulas using Excel and how to use Excel's drop-down menus that will create the formulas for you.
- This book does not come with a CD of Excel files which you can upload to your computer. Instead, you'll be shown how to create each Excel file yourself. In a work situation, your colleagues will not give you an Excel file; you will be expected to create your own. This book will give you ample practice in developing this important skill.
- Each chapter presents the steps needed to solve a practical science problem using Excel. In addition, there are three practice problems at the end of each chapter so you can test your new knowledge of statistics. The answers to these problems appear in Appendix A.
- A "Practice Test" is given in Appendix B to test your knowledge at the end of the book. The answers to these practical science problems appear in Appendix C.

Thomas Quirk, a current professor of marketing at the George Herbert Walker School of Business and Technology at Webster University in St. Louis, Missouri (USA), teaches marketing statistics, marketing research, and pricing strategies. He has published articles in the *Journal of Educational Psychology*, the *Journal of Educational Research*, the *Review of Educational Research*, the *Journal of Educational Measurement*, *Educational Technology*, the *Elementary School Journal*, the *Journal of Secondary Education*, *Educational Horizons*, and *Phi Delta Kappan*. In addition, Professor Quirk has written more than 60 textbook supplements in management and marketing, published more than 20 articles in professional journals, and presented more than 20 papers at professional meetings. He holds a B.S. in mathematics from John Carroll University, both an M.A. in education and a Ph.D. in educational psychology from Stanford University, and an M.B.A. from the University of Missouri-St. Louis.

Meghan Quirk holds both a Ph.D. in biological education and an M.A. in biological sciences from the University of Northern Colorado (UNC) and a B.A. in biology and religion at Principia College in Elsah, Illinois. She has done research on food web dynamics at Wind Cave National Park in South Dakota and research in agro-ecology in Southern Belize. She has co-authored an article on shortgrass steppe ecosystems in *Photochemistry and Photobiology* and has presented papers at the Shortgrass Steppe Symposium in Fort Collins, Colorado, and the Long-Term Ecological Research All Scientists Meeting in Estes Park, Colorado, and participated in the NSF Site Review of the Shortgrass Steppe Long-Term Ecological Research in Nunn, Colorado. She is a National Science Foundation Fellow GK-12 and currently teaches science in Bailey, Colorado.

Howard Horton holds an M.S. in biological sciences from the University of Northern Colorado (UNC) and a B.S. in biological sciences from Mesa State College. He has worked on research projects in Pawnee National Grasslands,

Rocky Mountain National Park; Long-Term Ecological Research at Toolik Lake, Alaska; and Wind Cave, South Dakota. He has co-authored articles in the *International Journal of Speleology* and the *Journal of Cave and Karst Studies*. He is a National Science Foundation Fellow GK-12 and is currently the angler outreach coordinator with the Colorado Division of Parks and Wildlife.

St. Louis, MO — Thomas J. Quirk
Bailey, CO — Meghan H. Quirk
Bailey, CO — Howard F. Horton

Acknowledgments

Excel 2016 for Biological and Life Sciences Statistics: A Guide to Solving Practical Problems is the result of inspiration from three important people: my two daughters and my wife. Jennifer Quirk McLaughlin invited me to visit her M.B.A. classes several times at the University of Witwatersrand in Johannesburg, South Africa. These visits to a first-rate M.B.A. program convinced me there was a need for a book to teach students how to solve practical problems using Excel. Meghan Quirk-Horton's dogged dedication to learning the many statistical techniques needed to complete her Ph.D. dissertation illustrated the need for a statistics book that would make this daunting task more user-friendly. And Lynne Buckley-Quirk was the number-one cheerleader for this project from the beginning, always encouraging me and helping me remain dedicated to completing it.

Thomas J. Quirk

We would like to acknowledge the patience of our two little girls, Lila and Elia, as we worked on this book with their TQ. We would also like to thank Professors Sarah Perkins, Doug Warren, John Moore, and Lee Dyer for their guidance and support during our college and graduate school careers.

Meghan H. Quirk and Howard F. Horton

Marc Strauss, our editor at Springer, caught the spirit of this idea in our first phone conversation and guided this book through the idea stages until it reached its final form. His encouragement and support, along with Christine Crigler's shepherding of this book through production, were vital to this book seeing the light of day. We thank them both for being such an outstanding product champion throughout this process.

Contents

1 Sample Size, Mean, Standard Deviation, and Standard Error of the Mean . . . 1
1.1 Mean . . . 1
1.2 Standard Deviation . . . 2
1.3 Standard Error of the Mean . . . 3
1.4 Sample Size, Mean, Standard Deviation, and Standard Error of the Mean . . . 4
1.4.1 Using the Fill/Series/Columns Commands . . . 4
1.4.2 Changing the Width of a Column . . . 5
1.4.3 Centering Information in a Range of Cells . . . 6
1.4.4 Naming a Range of Cells . . . 8
1.4.5 Finding the Sample Size Using the = COUNT Function . . . 9
1.4.6 Finding the Mean Score Using the =AVERAGE Function . . . 10
1.4.7 Finding the Standard Deviation Using the =STDEV Function . . . 10
1.4.8 Finding the Standard Error of the Mean . . . 10
1.5 Saving a Spreadsheet . . . 13
1.6 Printing a Spreadsheet . . . 14
1.7 Formatting Numbers in Currency Format (Two decimal places) . . . 15
1.8 Formatting Numbers in Number Format (Three Decimal Places) . . . 17
1.9 End-of-Chapter Practice Problems . . . 17
References . . . 20

2 Random Number Generator . . . 21
2.1 Creating Frame Numbers for Generating Random Numbers . . . 21
2.2 Creating Random Numbers in an Excel Worksheet . . . 24
2.3 Sorting Frame Numbers into a Random Sequence . . . 26

2.4 Printing an Excel File So That All of the Information Fits onto One Page 30
2.5 End-of-Chapter Practice Problems 33

3 Confidence Interval About the Mean Using the TINV Function and Hypothesis Testing 35
3.1 Confidence Interval About the Mean 35
3.1.1 How to Estimate the Population Mean 35
3.1.2 Estimating the Lower Limit and the Upper Limit of the 95 % Confidence Interval About the Mean 36
3.1.3 Estimating the Confidence Interval the Chevy Impala in Miles Per Gallon 37
3.1.4 Where Did the Number "1.96" Come From? 38
3.1.5 Finding the Value for t in the Confidence Interval Formula 39
3.1.6 Using Excel's TINV Function to Find the Confidence Interval About the Mean 40
3.1.7 Using Excel to find the 95 % Confidence Interval for a Car's mpg Claim 41
3.2 Hypothesis Testing 47
3.2.1 Hypotheses Always Refer to the Population of People, Plants, or Animals that You Are Studying 47
3.2.2 The Null Hypothesis and the Research (Alternative) Hypothesis 48
3.2.3 The 7 Steps for Hypothesis-Testing Using the Confidence Interval About the Mean 51
3.3 Alternative Ways to Summarize the Result of a Hypothesis Test 57
3.3.1 Different Ways to Accept the Null Hypothesis 58
3.3.2 Different Ways to Reject the Null Hypothesis 58
3.4 End-of-Chapter Practice Problems 59
References 63

4 One-Group t-Test for the Mean 65
4.1 The Seven STEPS for Hypothesis-Testing Using the One-Group t-Test 65
4.1.1 STEP 1: State the Null Hypothesis and the Research Hypothesis 66
4.1.2 STEP 2: Select the Appropriate Statistical Test 66
4.1.3 STEP 3: Decide on a Decision Rule for the One-Group t-Test 66
4.1.4 STEP 4: Calculate the Formula for the One-Group t-Test 67
4.1.5 STEP 5: Find the Critical Value of t in the t-Table in Appendix E 68

4.1.6 STEP 6: State the Result of Your Statistical Test 69
4.1.7 STEP 7: State the Conclusion of Your Statistical Test in Plain English! 69
4.2 One-Group t-Test for the Mean 70
4.3 Can You Use Either the 95 % Confidence Interval About the Mean OR the One-Group t-Test When Testing Hypotheses? 76
4.4 End-of-Chapter Practice Problems 76
References 80

5 Two-Group t-Test of the Difference of the Means for Independent Groups 81
5.1 The 9 STEPS for Hypothesis-Testing Using the Two-Group t-Test 82
5.1.1 STEP 1: Name One Group, Group 1, and the Other Group, Group 2 82
5.1.2 STEP 2: Create a Table That Summarizes the Sample Size, Mean Score, and Standard Deviation of Each Group 83
5.1.3 STEP 3: State the Null Hypothesis and the Research Hypothesis for the Two-Group t-Test 84
5.1.4 STEP 4: Select the Appropriate Statistical Test 84
5.1.5 STEP 5: Decide on a Decision Rule for the Two-Group t-Test 85
5.1.6 STEP 6: Calculate the Formula for the Two-Group t-Test 85
5.1.7 STEP 7: Find the Critical Value of t in the t-Table in Appendix E 85
5.1.8 STEP 8: State the Result of Your Statistical Test 86
5.1.9 STEP 9: State the Conclusion of Your Statistical Test in Plain English! 87
5.2 Formula #1: Both Groups Have a Sample Size Greater Than 30 91
5.2.1 An Example of Formula #1 for the Two-Group t-Test 92
5.3 Formula #2: One or Both Groups Have a Sample Size Less Than 30 98
5.4 End-of-Chapter Practice Problems 105
References 107

6 Correlation and Simple Linear Regression 109
6.1 What Is a "Correlation?" 109
6.1.1 Understanding the Formula for Computing a Correlation 114
6.1.2 Understanding the Nine Steps for Computing a Correlation, r 114

6.2 Using Excel to Compute a Correlation Between Two Variables 116
6.3 Creating a Chart and Drawing the Regression Line onto the Chart 121
6.3.1 Using Excel to Create a Chart and the Regression Line Through the Data Points 123
6.4 Printing a Spreadsheet So That the Table and Chart Fit Onto One Page 131
6.5 Finding the Regression Equation 133
6.5.1 Installing the Data Analysis ToolPak into Excel 134
6.5.2 Using Excel to Find the SUMMARY OUTPUT of Regression 137
6.5.3 Finding the Equation for the Regression Line 140
6.5.4 Using the Regression Line to Predict the y-Value for a Given x-Value 141
6.6 Adding the Regression Equation to the Chart 142
6.7 How to Recognize Negative Correlations in the SUMMARY OUTPUT Table 145
6.8 Printing Only Part of a Spreadsheet Instead of the Entire Spreadsheet 145
6.8.1 Printing Only the Table and the Chart on a Separate Page 146
6.8.2 Printing Only the Chart on a Separate Page 146
6.8.3 Printing Only the SUMMARY OUTPUT of the Regression Analysis on a Separate Page 147
6.9 End-of-Chapter Practice Problems 147
References 152

7 Multiple Correlation and Multiple Regression 153
7.1 Multiple Regression Equation 153
7.2 Finding the Multiple Correlation and the Multiple Regression Equation 156
7.3 Using the Regression Equation to Predict FRUIT PRODUCED 160
7.4 Using Excel to Create a Correlation Matrix in Multiple Regression 160
7.5 End-of-Chapter Practice Problems 164
References 169

8 One-Way Analysis of Variance (ANOVA) 171
8.1 Using Excel to Perform a One-Way Analysis of Variance (ANOVA) 173
8.2 How to Interpret the ANOVA Table Correctly 175
8.3 Using the Decision Rule for the ANOVA F-Test 176

8.4 Testing the Difference Between Two Groups Using the ANOVA t-Test 177
8.4.1 Comparing Subspecies B vs. Subspecies C in the Number of Sound Bursts per Cycle Using the ANOVA t-Test 177
8.4.1.1 Finding the Degrees of Freedom for the ANOVA t-test 178
8.4.1.2 Stating the Decision Rule for the ANOVA t-test 179
8.4.1.3 Performing an ANOVA t-Test Using Excel Commands 179
8.5 End-of-Chapter Practice Problems 182
References 188

Appendices 189
Appendix A: Answers to End-of-Chapter Practice Problems 189
Appendix B: Practice Test 223
Appendix C: Answers to Practice Test 233
Appendix D: Statistical Formulas 244
Appendix E: t-Table 246

Index 247

Chapter 1
Sample Size, Mean, Standard Deviation, and Standard Error of the Mean

This chapter deals with how you can use Excel to find the average (i.e., "mean") of a set of scores, the standard deviation of these scores (STDEV), and the standard error of the mean (s.e.) of these scores. All three of these statistics are used frequently and form the basis for additional statistical tests.

1.1 Mean

The *mean* is the "arithmetic average" of a set of scores. When my daughter was in the fifth grade, she came home from school with a sad face and said that she didn't get "averages." The book she was using described how to find the mean of a set of scores, and so I said to her:

"Jennifer, you add up all the scores and divide by the number of numbers that you have."

She gave me "that look," and said: "Dad, this is serious!" She thought I was teasing her. So I said:

"See these numbers in your book; add them up. What is the answer?" (She did that.)

"Now, how many numbers do you have?" (She answered that question.)

"Then, take the number you got when you added up the numbers, and divide that number by the number of numbers that you have."

She did that, and found the correct answer. You will use that same reasoning now, but it will be much easier for you because Excel will do all of the steps for you.

We will call this average of the scores the "mean" which we will symbolize as: $\overline{X}$, and we will pronounce it as: "Xbar."

The formula for finding the mean with your calculator looks like this:

$$\overline{X} = \frac{\Sigma X}{n} \tag{1.1}$$

T.J. Quirk et al., *Excel 2016 for Biological and Life Sciences Statistics*, Excel for Statistics, DOI 10.1007/978-3-319-39489-3_1

The symbol Σ is the Greek letter sigma, which stands for "sum." It tells you to add up all the scores that are indicated by the letter X, and then to divide your answer by n (the number of numbers that you have).

Let's give a simple example:

Suppose that you had these six biology test scores on an 7-item true-false quiz:

6
4
5
3
2
5

To find the mean of these scores, you add them up, and then divide by the number of scores. So, the mean is: 25/6 = 4.17

1.2 Standard Deviation

The *standard deviation* tells you "how close the scores are to the mean." If the standard deviation is a small number, this tells you that the scores are "bunched together" close to the mean. If the standard deviation is a large number, this tells you that the scores are "spread out" a greater distance from the mean. The formula for the standard deviation (which we will call STDEV) and use the letter, S, to symbolize is:

$$\text{STDEV} = S = \sqrt{\frac{\Sigma(X - \overline{X})^2}{n - 1}} \tag{1.2}$$

The formula look complicated, but what it asks you to do is this:

1. Subtract the mean from each score $(X - \overline{X})$.
2. Then, square the resulting number to make it a positive number.
3. Then, add up these squared numbers to get a total score.
4. Then, take this total score and divide it by n − 1 (where n stands for the number of numbers that you have).
5. The final step is to take the square root of the number you found in step 4.

You will not be asked to compute the standard deviation using your calculator in this book, but you could see examples of how it is computed in any basic statistics book. Instead, we will use Excel to find the standard deviation of a set of scores. When we use Excel on the six numbers we gave in the description of the mean above, you will find that the *STDEV* of these numbers, S, is 1.47.

1.3 Standard Error of the Mean

The formula for the *standard error of the mean* (*s.e.*, which we will use $S_{\overline{X}}$ to symbolize) is:

$$\text{s.e.} = S_{\overline{X}} = \frac{S}{\sqrt{n}} \tag{1.3}$$

To find ***s.e.***, all you need to do is to take the standard deviation, STDEV, and divide it by the square root of n, where n stands for the "number of numbers" that you have in your data set. In the example under the standard deviation description above, the $s.e. = 0.60$. (You can check this on your calculator.)

If you want to learn more about the standard deviation and the standard error of the mean, see Bremer and Doerge (2010) and Weiers (2011).

Now, let's learn how to use Excel to find the sample size, the mean, the standard deviation, and the standard error or the mean using the weight (measured in grams) of eight adult wild field mice collected in a tall grass prairie on the bluffs above the Mississippi River near St. Louis, Missouri. The hypothetical data appear in Fig. 1.1.

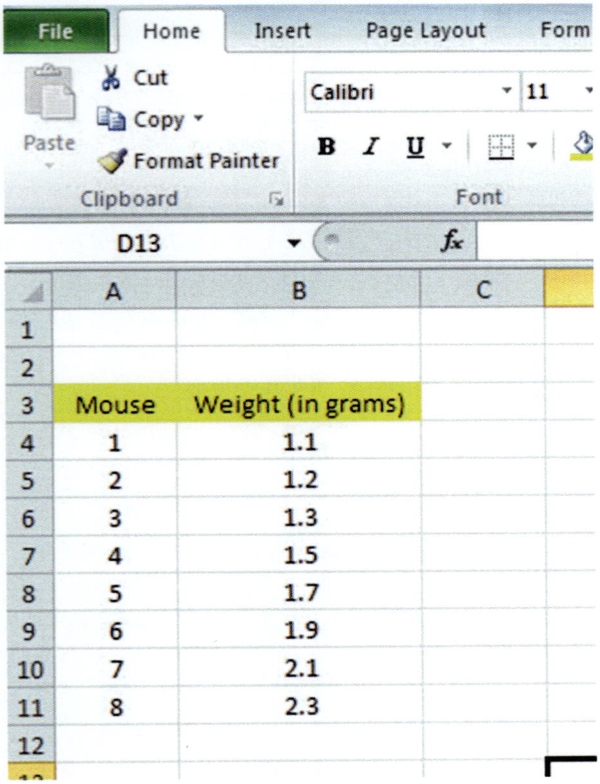

Fig. 1.1 Worksheet Data for Wild Field Mice Weights (Practical Example)

1.4 Sample Size, Mean, Standard Deviation, and Standard Error of the Mean

Objective: To find the sample size (n), mean, standard deviation (STDEV), and standard error of the mean (s.e.) for these data

Start your computer, and click on the Excel 2016 icon to open a blank Excel spreadsheet.

Click on: Blank workbook

Enter the data in this way:

A3: Mouse
B3: Weight (in grams)
A4: 1

1.4.1 Using the Fill/Series/Columns Commands

Objective: To add the mouse numbers 2–8 in a column underneath Mouse #1

Put pointer in A4
Home (top left of screen)

Important note: The "Paste" command should be on the top of your screen on the far left of the screen.

Important note: Notice the Excel commands at the top of your computer screen: File Home Insert Page Layout Formulas etc.
If these commands ever "disappear" when you are using Excel, you need to click on "Home" at the top left of your screen to make them reappear!

Fill (top right of screen: click on the down arrow; see Fig. 1.2)

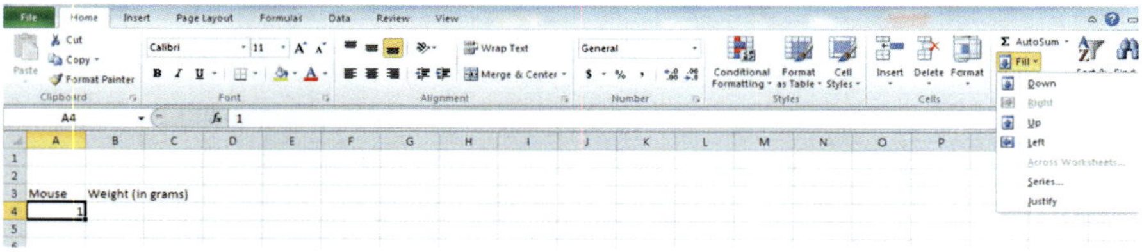

Fig. 1.2 Home/Fill/Series commands

Series
Columns
Step value: 1
Stop value: 8 (see Fig. 1.3)

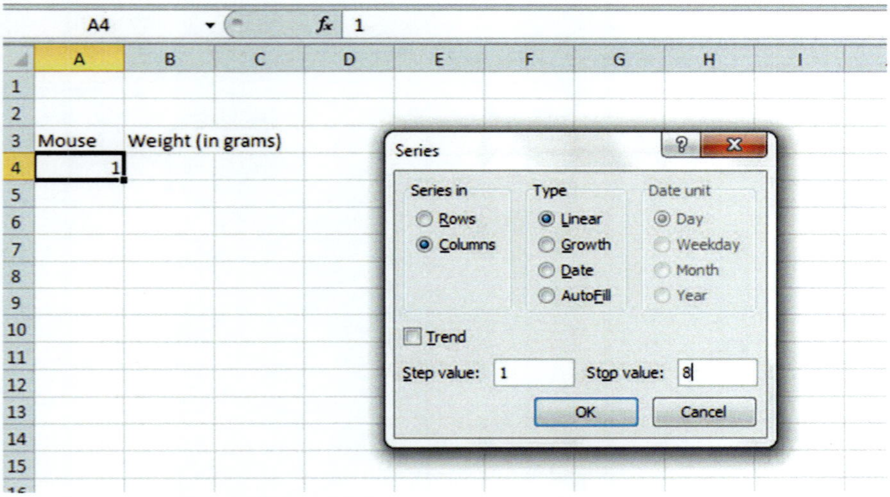

Fig. 1.3 Example of Dialogue Box for Fill/Series/Columns/Step Value/Stop Value commands

OK

The mice numbers should be identified as 1–8, with 8 in cell A11.

Now, enter the mice weights in cells B4: B11. *(Note: Be sure to double-check your figures to make sure that they are correct or you will not get the correct answer!)*

Since your computer screen shows the information in a format that does not look professional, you need to learn how to "widen the column width" and how to "center the information" in a group of cells. Here is how you can do those two steps:

1.4.2 Changing the Width of a Column

Objective: To make a column width wider so that all of the information fits inside that column

If you look at your computer screen, you can see that Column B is not wide enough so that all of the information fits inside this column. To make Column B wider:

Click on the letter, B, at the top of your computer screen

Place your mouse pointer on your computer at the far right corner of B until you create a "cross sign" on that corner

Left-click on your mouse, hold it down, and move this corner to the right until it is "wide enough to fit all of the data"

Take your finger off the mouse to set the new column width (see Fig. 1.4)

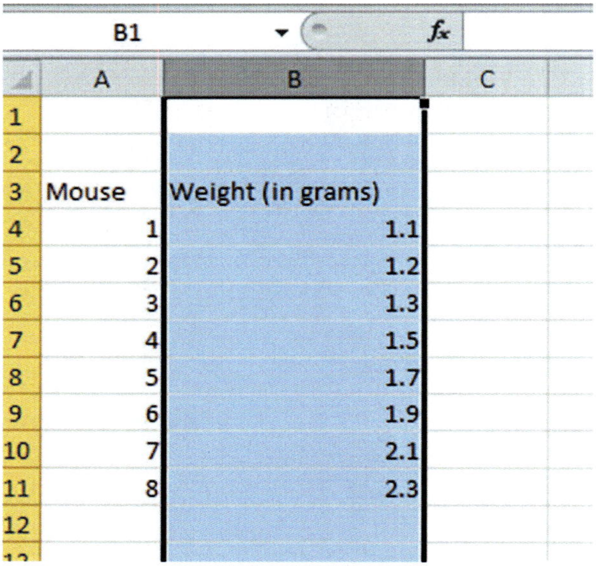

Fig. 1.4 Example of How to Widen the Column Width

Then, click on any empty cell (i.e., any blank cell) to "deselect" column B so that it is no longer a darker color on your screen.

When you widen a column, you will make all of the cells in all of the rows of this column that same width.

Now, let's go through the steps to center the information in both Column A and Column B.

1.4.3 Centering Information in a Range of Cells

Objective: To center the information in a group of cells

In order to make the information in the cells look "more professional," you can center the information using the following steps:

Left-click your mouse pointer on A3 and drag it to the right and down to highlight cells A3:B11 so that these cells appear in a darker color

Home

At the top of your computer screen, you will see a set of "lines" in which all of the lines are "centered" to the same width under "Alignment" (it is the second icon at the bottom left of the Alignment box; see Fig. 1.5)

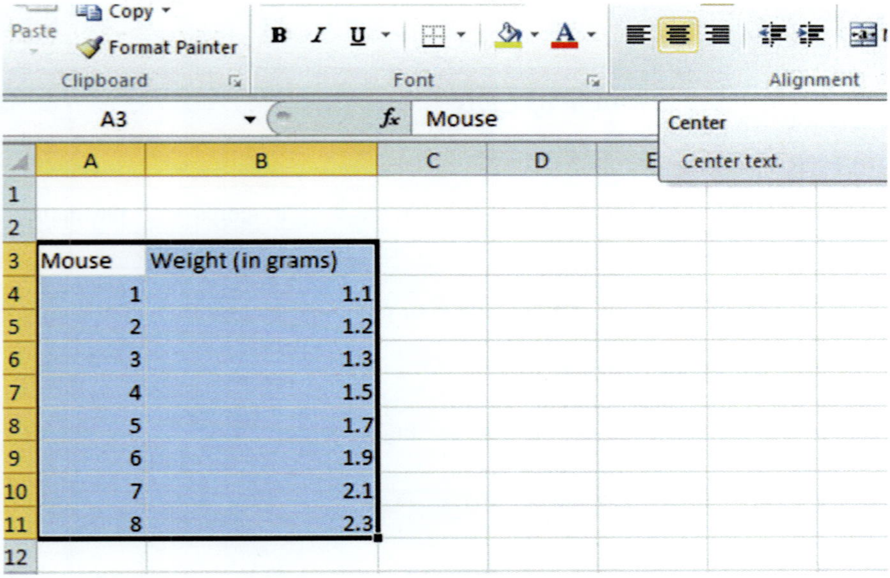

Fig. 1.5 Example of How to Center Information Within Cells

Click on this icon to center the information in the selected cells (see Fig. 1.6)

Fig. 1.6 Final Result of Centering Information in the Cells

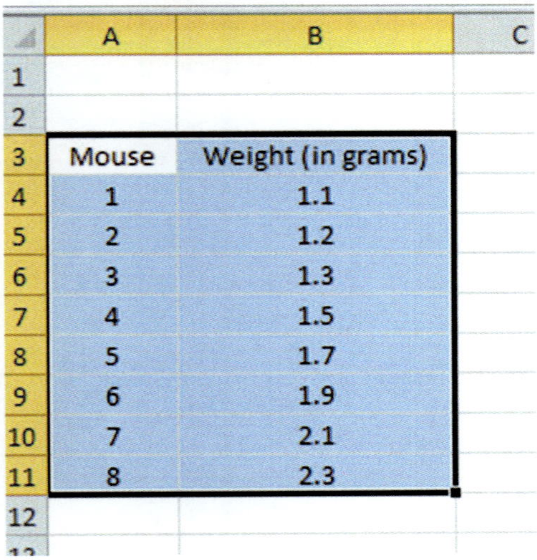

Since you will need to refer to the mice weights in your formulas, it will be much easier to do this if you "name the range of data" with a name instead of having to remember the exact cells (B4: B11) in which these figures are located. Let's call that group of cells: Weight, but we could give them any name that you want to use.

1.4.4 *Naming a Range of Cells*

Objective: To name the range of data for the weight scores with the name: Weight

Highlight cells B4: B11 by left-clicking your mouse pointer on B4 and dragging it down to B11

Formulas (top left of your screen)

Define Name (top center of your screen)

Weight (type this name in the top box; see Fig. 1.7)

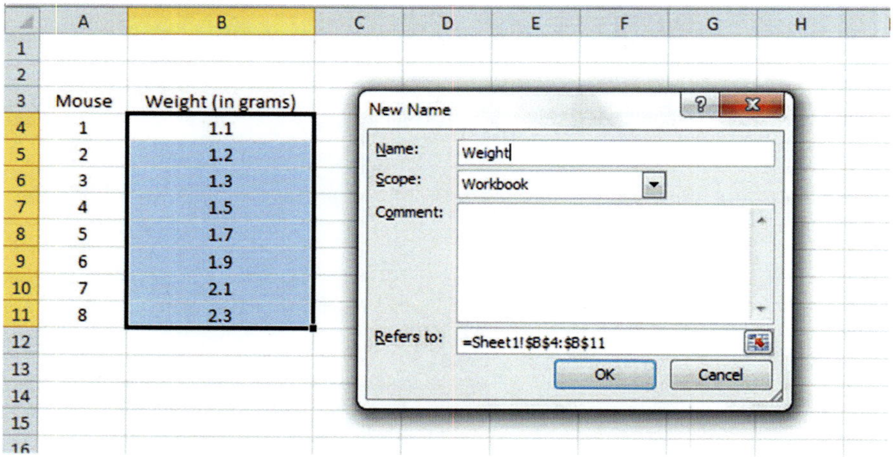

Fig. 1.7 Dialogue box for "naming a range of cells" with the name: Weight

OK

Then, click on any cell of your spreadsheet that does not have any information in it (i.e., it is an "empty cell") to deselect cells B4:B11

Now, add the following terms to your spreadsheet:

E6: n
E9: Mean
E12: STDEV
E15: s.e. (see Fig. 1.8)

	A	B	C	D	E	F
1						
2						
3	Mouse	Weight (in grams)				
4	1	1.1				
5	2	1.2				
6	3	1.3			n	
7	4	1.5				
8	5	1.7				
9	6	1.9			Mean	
10	7	2.1				
11	8	2.3				
12					STDEV	
13						
14						
15					s.e.	
16						

Fig. 1.8 Example of Entering the Sample Size, Mean, STDEV, and s.e. Labels

Note: Whenever you use a formula, you must add an equal sign (=) at the beginning of the name of the function so that Excel knows that you intend to use a formula.

1.4.5 Finding the Sample Size Using the =COUNT Function

> Objective: To find the sample size (n) for these data using the =COUNT function

F6: =COUNT(Weight)

This command should insert the number 8 into cell F6 since there are eight mice in your sample.

1.4.6 Finding the Mean Score Using the =AVERAGE Function

Objective: To find the mean weight figure using the =AVERAGE function

F9: =AVERAGE(Weight)

This command should insert the number 1.6375 into cell F9.

1.4.7 Finding the Standard Deviation Using the =STDEV Function

Objective: To find the standard deviation (STDEV) using the =STDEV function

F12: =STDEV(Weight)

This command should insert the number 0.437321 into cell F12.

1.4.8 Finding the Standard Error of the Mean

Objective: To find the standard error of the mean using a formula for these eight data points

F15: =F12/SQRT(8)

This command should insert the number 0.154616 into cell F15 (see Fig. 1.9).

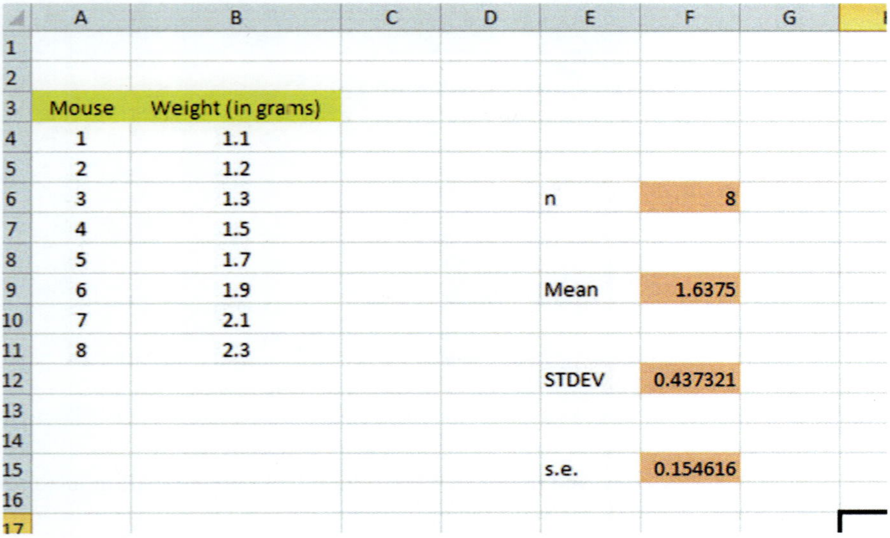

	A	B	C	D	E	F	G
1							
2							
3	Mouse	Weight (in grams)					
4	1	1.1					
5	2	1.2					
6	3	1.3			n	8	
7	4	1.5					
8	5	1.7					
9	6	1.9			Mean	1.6375	
10	7	2.1					
11	8	2.3					
12					STDEV	0.437321	
13							
14							
15					s.e.	0.154616	
16							
17							

Fig. 1.9 Example of Using Excel Formulas for Sample Size, Mean, STDEV, and s.e.

Important note: Throughout this book, be sure to double-check all of the figures in your spreadsheet to make sure that they are in the correct cells, or the formulas will not work correctly!

1.4.8.1 Formatting Numbers in Number Format (Two decimal places)

Objective: To convert the mean, STDEV, and s.e. to two decimal places

Highlight cells F9:F15

Home (top left of screen)

Look under "Number" at the top center of your screen. In the bottom right corner, gently place your mouse pointer on you screen at the bottom of the .00 .0 until it says: "Decrease Decimal" (see Fig. 1.10)

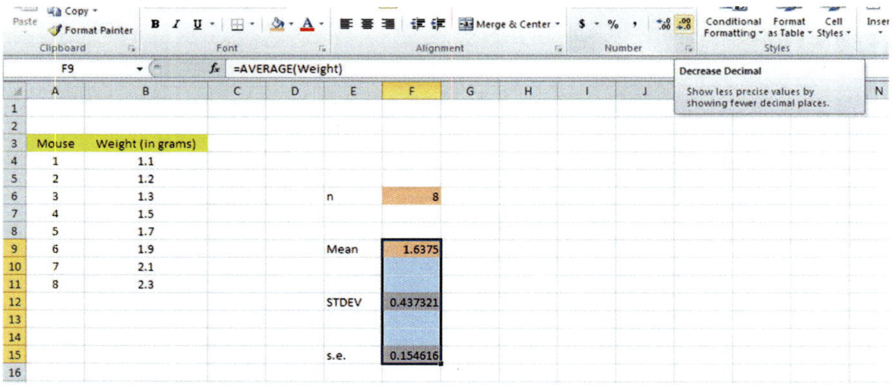

Fig. 1.10 Using the "Decrease Decimal Icon" to convert Numbers to Fewer Decimal Places

Click on this icon *twice* and notice that the cells F9:F15 are now all in just two decimal places (see Fig. 1.11)

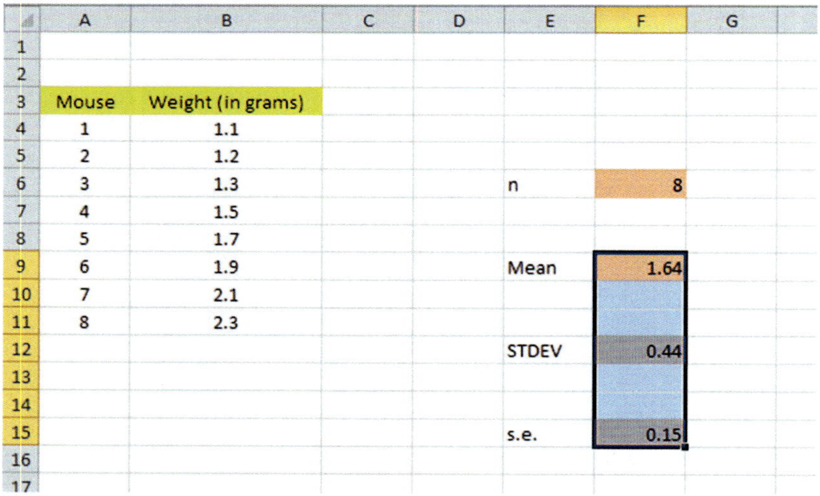

Fig. 1.11 Example of Converting Numbers to Two Decimal Places

Now, click on any "empty cell" on your spreadsheet to deselect cells F9:F15.

1.5 Saving a Spreadsheet

Objective: To save this spreadsheet with the name: Weight3

In order to save your spreadsheet so that you can retrieve it sometime in the future, your first decision is to decide "where" you want to save it. That is your decision and you have several choices. If it is your own computer, you can save it onto your hard drive (you need to ask someone how to do that on your computer). Or, you can save it onto a "CD" or onto a "flash drive." You then need to complete these steps:

File

Save as

(select the place where you want to save the file by scrolling either down or up the bar on the left, and click on the place where you want to save the file; for example: This PC: My Documents location)

File name: Weight3 (enter this name to the right of File name; see Fig. 1.12)

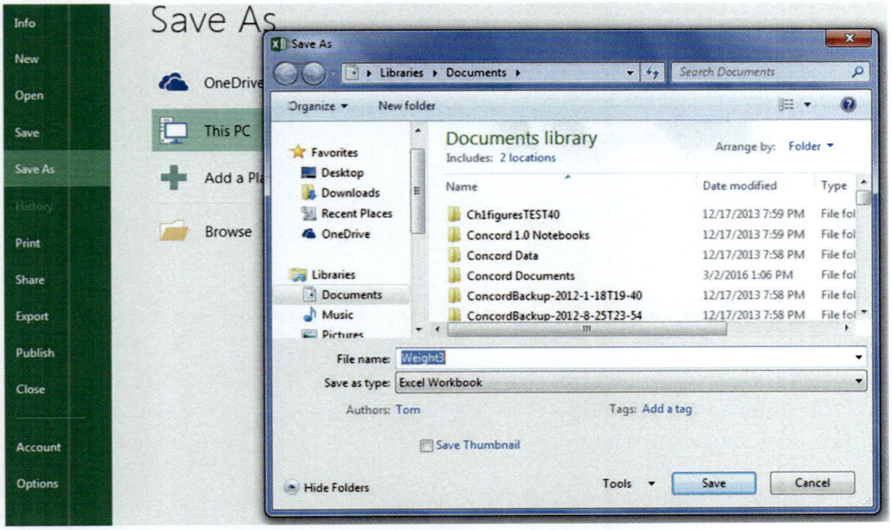

Fig. 1.12 Dialogue Box of Saving an Excel Workbook File as "Weight3" in My Documents location

Save

Important note: Be very careful to save your Excel file spreadsheet every few minutes so that you do not lose your information!

1.6 Printing a Spreadsheet

Objective: To print the spreadsheet

Use the following procedure when printing any spreadsheet.
File
Print
Print Active Sheets (see Fig. 1.13)

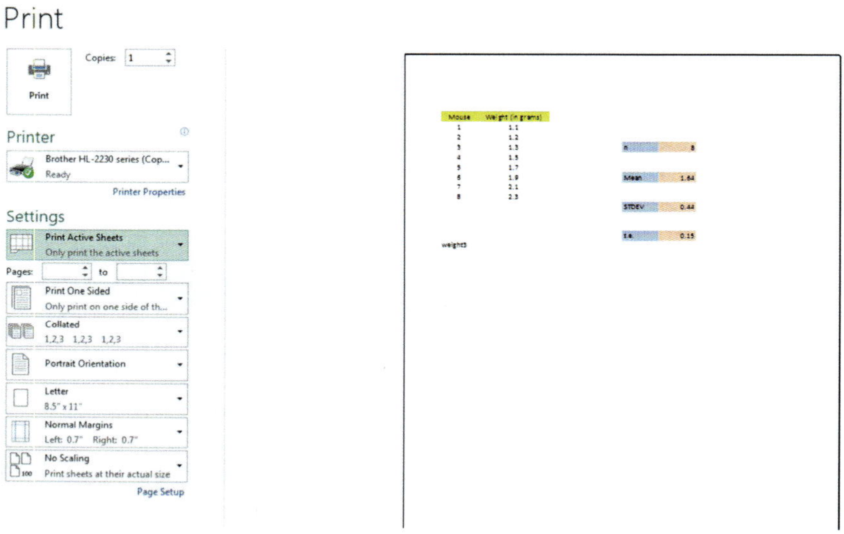

Fig. 1.13 Example of How to Print an Excel Worksheet Using the File/Print/Print Active Sheets Commands

Print (top of your screen)
The final spreadsheet is given in Fig 1.14

	A	B	C	D	E	F	G	H
1								
2								
3	Mouse	Weight (in grams)						
4	1	1.1						
5	2	1.2						
6	3	1.3			n	8		
7	4	1.5						
8	5	1.7						
9	6	1.9			Mean	1.64		
10	7	2.1						
11	8	2.3						
12					STDEV	0.44		
13								
14								
15					s.e.	0.15		
16								

Fig. 1.14 Final Result of Printing an Excel Spreadsheet

Before you leave this chapter, let's practice changing the format of the figures on a spreadsheet with two examples: (1) using two decimal places for figures that are dollar amounts, and (2) using three decimal places for figures.

Save the final spreadsheet by: File/Save, then close your spreadsheet by: File/Close, and open a blank Excel spreadsheet by using File/New/Blank Workbook (on the far left of your screen).

1.7 Formatting Numbers in Currency Format (Two decimal places)

Objective: To change the format of figures to dollar format with two decimal places

A3: Price
A4: 1.25
A5: 3.45
A6: 12.95

Highlight cells A4:A6 by left-clicking your mouse on A4 and dragging it down so that these three cells are highlighted in a darker color
Home
Number (top center of screen: click on the down arrow on the right; see Fig. 1.15)

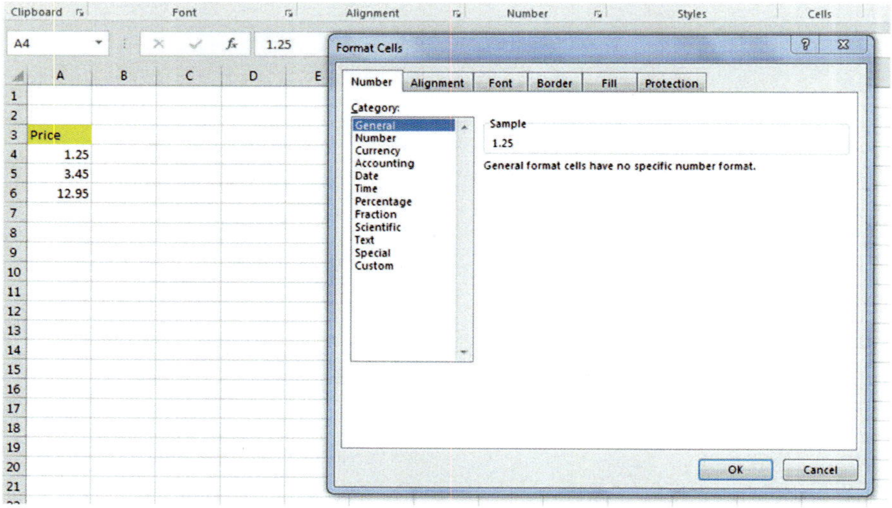

Fig. 1.15 Dialogue Box for Number Format Choices

Category: Currency
Decimal places: 2 (then see Fig. 1.16)

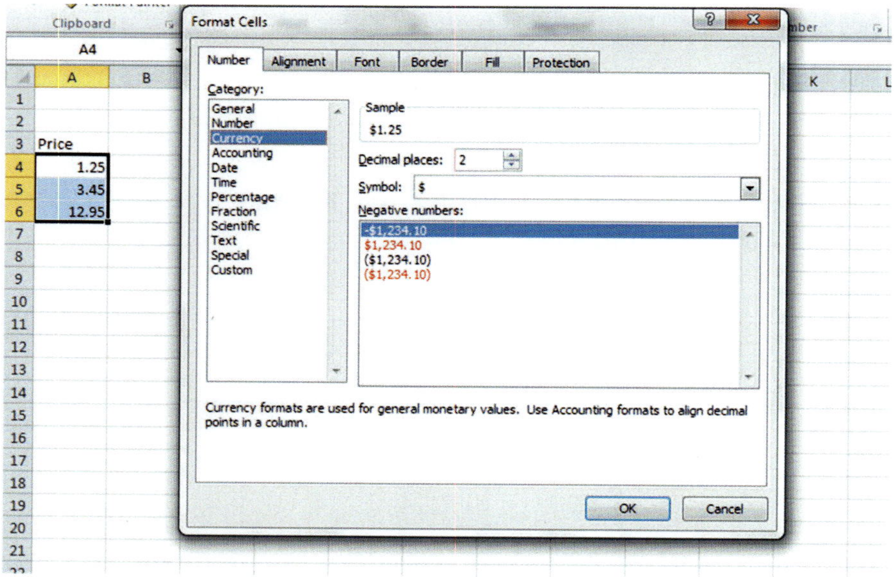

Fig. 1.16 Dialogue Box for Currency (two decimal places) Format for Numbers

OK

The three cells should have a dollar sign in them and be in two decimal places. Next, let's practice formatting figures in number format, three decimal places.

1.8 Formatting Numbers in Number Format (Three Decimal Places)

Objective: To format figures in number format, three decimal places

Home
Highlight cells A4:A6 on your computer screen
Number (click on the down arrow on the right)
Category: number
At the right of the box, change two decimal places to three decimal places by clicking on the "up arrow" once

OK

The three figures should now be in number format, each with three decimals.

Now, click on any blank cell to deselect cells A4:A6. Then, close this file by File/Close/Don't Save (since there is no need to save this practice problem).

You can use these same commands to format a range of cells in percentage format (and many other formats) to whatever number of decimal places you want to specify.

1.9 End-of-Chapter Practice Problems

1. Suppose that you wanted to find the mean, standard deviation, and standard error of the mean for the number of weed (*Potentilla*) seeds in a sample of grass seeds (*Phleum pratense*) as measured by the total number of seeds in a quarter-ounce sample of grass seeds. This interesting problem is discussed further in Sokal and Rohlf (1969). The hypothetical data appear in Fig. 1.17.

Number of weed seeds in a sample of grass seeds
No. of seeds
1
3
2
0
4
6
5
7
0
2
3
4
2
3
1
3
4

Fig. 1.17 Worksheet Data for Chap. 1: Practice Problem #1

(a) Use Excel to the right of the table to find the sample size, mean, standard deviation, and standard error of the mean for these data. Label your answers, and round off the mean, standard deviation, and standard error of the mean to two decimal places; use number format for these three figures.
(b) Print the result on a separate page.
(c) Save the file as: seed3

2. Suppose that you have been hired as a research assistant and that you have been asked to determine the percent of butterfat in registered 3-year-old Ayrshire cows in the state of Montana in the USA. You manage to find a recent "stock record book" with these data for this breed of cow in Montana, and you take a random sample of cows and record the percent of butterfat of each of these cows from the stock record book. The hypothetical data are given in Fig. 1.18.

REGISTERED THREE-YEAR-OLD AYRSHIRE COWS IN MONTANA
PERCENT OF BUTTERFAT
3.70
5.10
3.95
3.84
3.92
4.10
4.32
4.84
4.61
5.00
4.92
3.86
3.97
4.23
4.54
4.89
4.23
4.15

Fig. 1.18 Worksheet Data for Chap. 1: Practice Problem #2

(a) Use Excel to create a table of these data, and at the right of the table use Excel to find the sample size, mean, standard deviation, and standard error of the mean for these data. Label your answers, and round off the mean, standard deviation, and standard error of the mean to two decimal places using number format.
(b) Print the result on a separate page.
(c) Save the file as: COWS3

3. Suppose that a 5th grade elementary science teacher in Fairbanks, Alaska, is using a textbook based on human anatomy that typically requires about eight class days to teach each chapter. At the end of Chap. 8, the teacher gives a 15-item true-false quiz on this chapter. The test results are given in Fig. 1.19:

Fig. 1.19 Worksheet Data for Chap. 1: Practice Problem #3

Fairbanks, Alaska, Elementary School
5th grade anatomy science test
Chapter 8 (15 items)
12
15
13
8
10
12
13
12
9
4
11
15
13
15
12
14

(a) Use Excel to create a table for these data, and at the right of the table, use Excel to find the sample size, mean, standard deviation, and standard error of the mean for these data. Label your answers, and round off the mean, standard deviation, and standard error of the mean to three decimal places using number format.
(b) Print the result on a separate page.
(c) Save the file as: SCIENCE9

References

Bremer, M. and Doerge, R.W. Statistics at the Bench: A Step-by-Step Handbook for Biologists. Cold Spring Harbor, NY: Cold Spring Harbor Laboratory Press, 2010.

Sokal, R.R. and Rohlf, F.J. Biometry: The Principles and Practice of Statistics in Biological Research. San Francisco, CA: W.H. Freeman and Company, 1969.

Weiers, R.M. Introduction to Business Statistics (7th ed.). Mason, OH: South-Western Cengage Learning, 2011.

Chapter 2
Random Number Generator

Suppose that a local poultry farmer is raising a variety of laying hens and that he is interested in the egg production of a particular breed of hens (Breed X). Suppose that he has asked you to take a random sample of 5 of Breed X's 32 laying hens to identify overall egg production of this breed. Using your Excel skills to determine egg production, you will need to define a "sampling frame." A sampling frame is a list of objects, events, or people from which you want to select a random sample. In this case, it is the group of 32 chickens. The frame starts with the identification code (ID) of the number 1 that is assigned to the first hen in the group of 32 laying hens on the poultry farm. The second hen has a code number of 2, the third a code number of 3, and so forth until the last hen has a code number of 32.

Since the poultry farm has 32 Breed X hens, your sampling frame would go from 1 to 32 with each hen having a unique ID number.

We will first create the frame numbers as follows in a new Excel worksheet:

2.1 Creating Frame Numbers for Generating Random Numbers

Objective: To create the frame numbers for generating random numbers

A3: FRAME NO.
A4: 1

Now, create the frame numbers in column A with the Home/Fill commands that were explained in the first chapter of this book (see Sect. 1.4.1) so that the frame numbers go from 1 to 32, with the number 32 in cell A35. If you need to be reminded about how to do that, here are the steps:

T.J. Quirk et al., *Excel 2016 for Biological and Life Sciences Statistics*, Excel for Statistics, DOI 10.1007/978-3-319-39489-3_2

Click on cell A4 to select this cell
Home
Fill (then click on the "down arrow" next to this command and select)
Series (see Fig. 2.1)

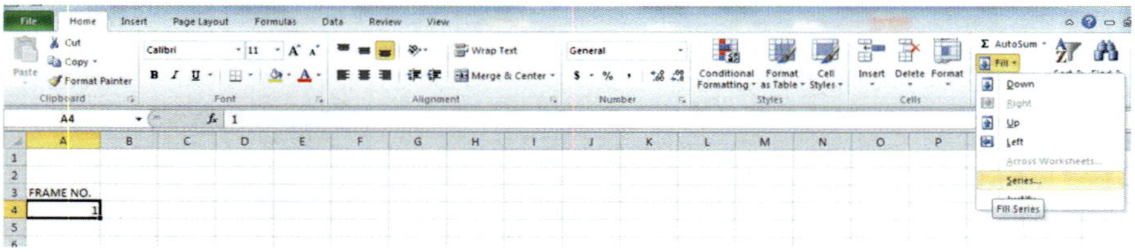

Fig. 2.1 Dialogue Box for Fill/Series Commands

Columns
Step value: 1
Stop value: 32 (see Fig. 2.2)

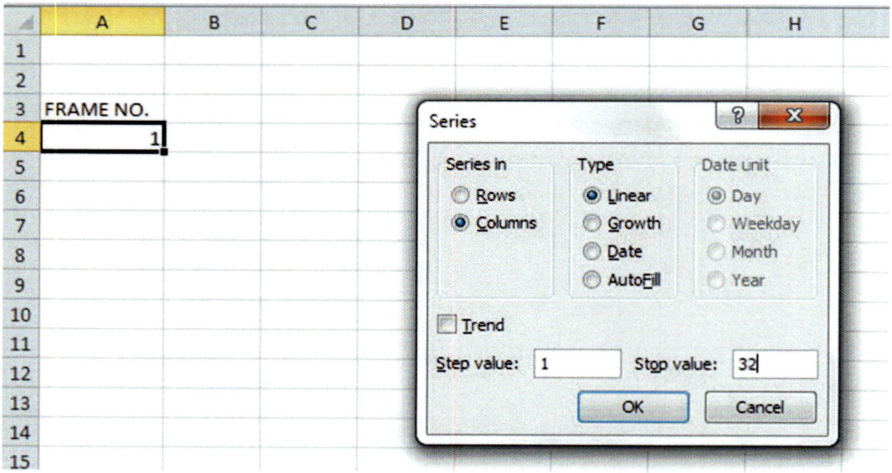

Fig. 2.2 Dialogue Box for Fill/Series/Columns/Step value/ Stop value Commands

OK

Then, save this file as: Random29. You should obtain the result in Fig. 2.3.

Fig. 2.3 Frame Numbers from 1 to 32

FRAME NO.
1
2
3
4
5
6
7
8
9
10
11
12
13
14
15
16
17
18
19
20
21
22
23
24
25
26
27
28
29
30
31
32

Now, create a column next to these frame numbers in this manner:

B3: DUPLICATE FRAME NO.
B4: 1

Next, use the Home/Fill command again, so that the 32 frame numbers begin in cell B4 and end in cell B35. Be sure to widen the columns A and B so that all of the information in these columns fits inside the column width. Then, center the information inside both Column A and Column B on your spreadsheet. You should obtain the information given in Fig. 2.4.

Fig. 2.4 Duplicate Frame Numbers from 1 to 32

FRAME NO.	DUPLICATE FRAME NO.
1	1
2	2
3	3
4	4
5	5
6	6
7	7
8	8
9	9
10	10
11	11
12	12
13	13
14	14
15	15
16	16
17	17
18	18
19	19
20	20
21	21
22	22
23	23
24	24
25	25
26	26
27	27
28	28
29	29
30	30
31	31
32	32

Save this file as: Random30

You are probably wondering why you created the same information in both Column A and Column B of your spreadsheet. This is to make sure that before you sort the frame numbers that you have exactly 32 of them when you finish sorting them into a random sequence of 32 numbers.

Now, let's add a random number to each of the duplicate frame numbers as follows:

2.2 Creating Random Numbers in an Excel Worksheet

C3: RANDOM NO.
(then widen columns A, B, C so that their labels fit inside the columns; then center the information in A3:C35)

C4: =RAND()

Next, hit the Enter key to add a random number to cell C4.

Note that you need *both* an open parenthesis *and* a closed parenthesis after = *RAND*(). The RAND command "looks to the left of the cell with the RAND() COMMAND in it" and assigns a random number to that cell.

Now, put the pointer using your mouse in cell C4 and then move the pointer to the bottom right corner of that cell until you see a "plus sign" in that cell. Then, click and drag the pointer down to cell C35 to add a random number to all 32 ID frame numbers (see Fig. 2.5).

Fig. 2.5 Example of Random Numbers Assigned to the Duplicate Frame Numbers

FRAME NO.	DUPLICATE FRAME NO.	RANDOM NO.
1	1	0.178997426
2	2	0.269196787
3	3	0.48649709
4	4	0.882904516
5	5	0.015953504
6	6	0.099651545
7	7	0.42850057
8	8	0.381659988
9	9	0.431296832
10	10	0.476642453
11	11	0.268603728
12	12	0.871330234
13	13	0.775421903
14	14	0.908450998
15	15	0.138749452
16	16	0.159535582
17	17	0.672417279
18	18	0.956231064
19	19	0.486746795
20	20	0.83596565
21	21	0.688574546
22	22	0.467838617
23	23	0.695493167
24	24	0.226521237
25	25	0.335451708
26	26	0.209245145
27	27	0.631291464
28	28	0.210229448
29	29	0.553196562
30	30	0.494647331
31	31	0.986702143
32	32	0.178067956

Then, click on any empty cell to deselect C4:C35 to remove the dark color highlighting these cells.

Save this file as: Random31

Now, let's sort these duplicate frame numbers into a random sequence:

2.3 Sorting Frame Numbers into a Random Sequence

Objective: To sort the duplicate frame numbers into a random sequence

Highlight cells B3 : C35 (include the labels at the top of columns B and C)
Data (top of screen)
Sort (click on this word at the top center of your screen; see Fig. 2.6)

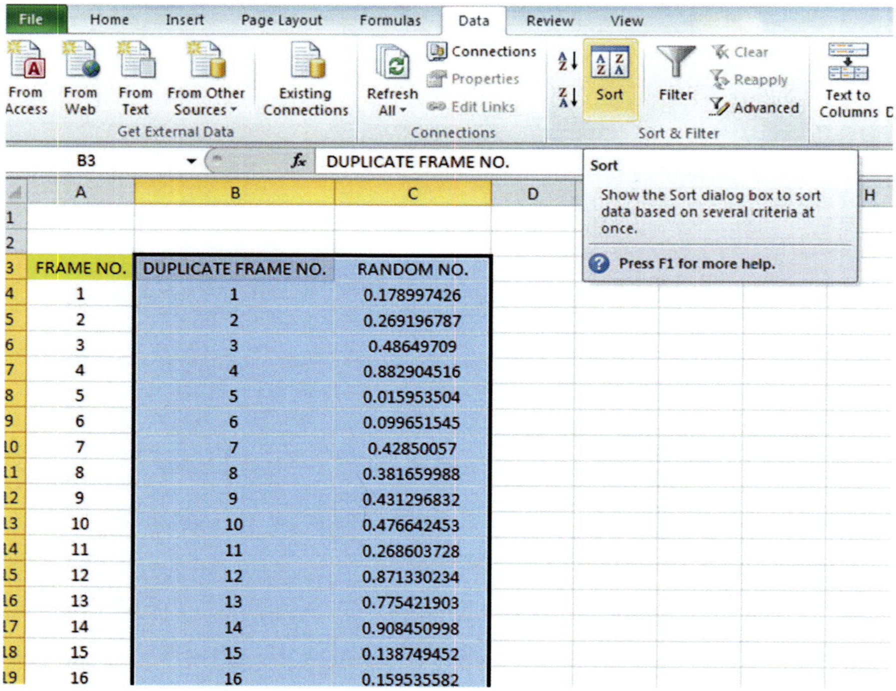

Fig. 2.6 Dialogue Box for Data/Sort Commands

Sort by: RANDOM NO. (click on the down arrow)
Smallest to Largest (see Fig. 2.7)

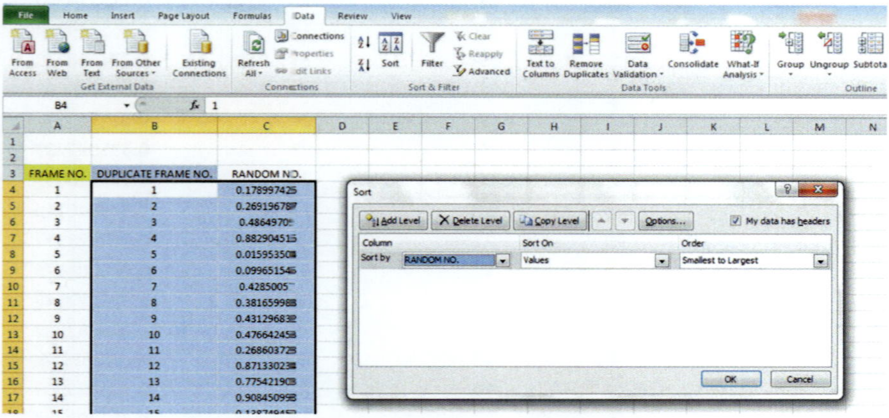

Fig. 2.7 Dialogue Box for Data/Sort/RANDOM NO./Smallest to Largest Commands

OK
Click on any empty cell to deselect B3:C35.
Save this file as: Random32
Print this file now.

These steps will produce Fig. 2.8 with the DUPLICATE FRAME NUMBERS sorted into a random order.

Important note: Because Excel randomly assigns these random numbers, your Excel commands will produce a different sequence of random numbers from everyone else who reads this book!

Fig. 2.8 Duplicate Frame Numbers Sorted by Random Number

FRAME NO.	DUPLICATE FRAME NO.	RANDOM NO.
1	5	0.063981403
2	6	0.977468743
3	15	0.225170263
4	16	0.765734052
5	32	0.274680922
6	1	0.594468001
7	26	0.511966171
8	28	0.625577233
9	24	0.906310053
10	11	0.488640116
11	2	0.020129977
12	25	0.723003676
13	8	0.975227547
14	7	0.469582962
15	9	0.14889954
16	22	0.955629903
17	10	0.897398234
18	3	0.314860892
19	19	0.442019486
20	30	0.078566335
21	29	0.172474705
22	27	0.104689528
23	17	0.406630369
24	21	0.961398315
25	23	0.094222677
26	13	0.323429051
27	20	0.470615753
28	12	0.978014724
29	4	0.618082813
30	14	0.727776384
31	18	0.919475329
32	31	0.324497007

Because your objective at the beginning of this chapter was to select randomly 5 of Breed X's 32 laying hens, you now can do that by selecting the *first five ID numbers* in DUPLICATE FRAME NO. column after the sort.

Although your first five random numbers will be different from those we have selected in the random sort that we did in this chapter, we would select these five IDs of hens using Fig. 2.9.

5, 6, 15, 16, 32

Fig. 2.9 First Five Hens Selected Randomly

FRAME NO.	DUPLICATE FRAME NO.	RANDOM NO.
1	5	0.063981403
2	6	0.977468743
3	15	0.225170263
4	16	0.765734052
5	32	0.274680922
6	1	0.594468001
7	26	0.511966171
8	28	0.625577233
9	24	0.906310053
10	11	0.488640116
11	2	0.020129977
12	25	0.723003676
13	8	0.975227547
14	7	0.469582962
15	9	0.14889954
16	22	0.955629903
17	10	0.897398234
18	3	0.314860892
19	19	0.442019486
20	30	0.078566335
21	29	0.172474705
22	27	0.104689528
23	17	0.406630369
24	21	0.961398315
25	23	0.094222677
26	13	0.323429051
27	20	0.470615753
28	12	0.978014724
29	4	0.618082813
30	14	0.727776384
31	18	0.919475329
32	31	0.324497007

Save this file as: Random33

Remember, your five ID numbers selected after your random sort will be different from the five ID numbers in Fig. 2.9 because Excel assigns a different random number *each time the =RAND() command is given.*

Before we leave this chapter, you need to learn how to print a file so that all of the information on that file fits onto a single page without "dribbling over" onto a second or third page.

2.4 Printing an Excel File So That All of the Information Fits onto One Page

Objective: To print a file so that all of the information fits onto one page

Note that the three practice problems at the end of this chapter require you to sort random numbers when the files contain 63 test tubes, 114 steel samples, and 75 field mice, respectively. These files will be "too big" to fit onto one page when you print them unless you format these files so that they fit onto a single page when you print them.

Let's create a situation where the file does not fit onto one printed page unless you format it first to do that.

Go back to the file you just created, Random 33, and enter the name: *Jennifer* into cell: A50.

If you printed this file now, the name, *Jennifer*, would be printed onto a second page because it "dribbles over" outside of the page range for this file in its current format.

So, you would need to change the page format so that all of the information, including the name, Jennifer, fits onto just one page when you print this file by using the following steps:

Page Layout (top left of the computer screen)
(Notice the "Scale to Fit" section in the center of your screen; see Fig. 2.10)

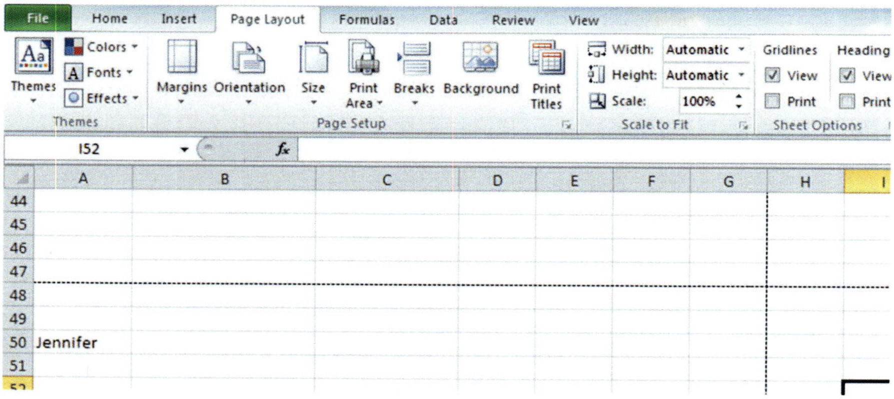

Fig. 2.10 Dialogue Box for Page Layout/Scale to Fit Commands

Hit the down arrow to the right of 100 % *once* to reduce the size of the page to 95 %.

Now, note that the name, Jennifer, is still on a second page on your screen because her name is below the horizontal dotted line on your screen in Fig. 2.11 (the dotted lines tell you outline dimensions of the file if you printed it now).

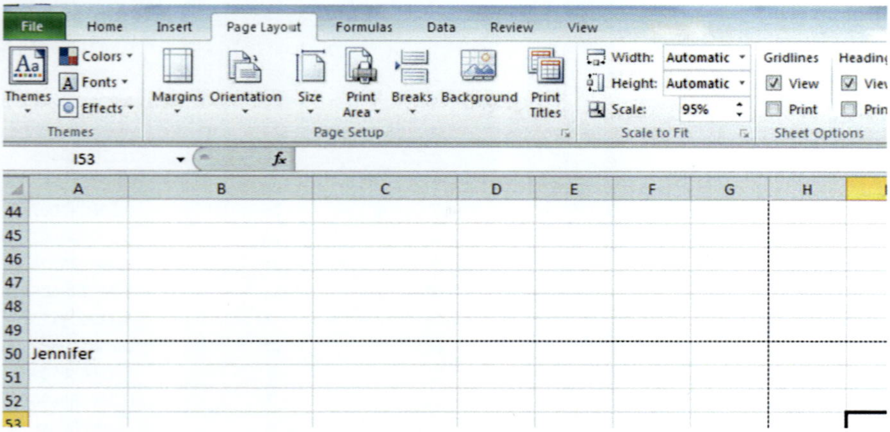

Fig. 2.11 Example of Scale Reduced to 95 % with "Jennifer" to be Printed on a Second Page

So, you need to repeat the "scale change steps" by hitting the down arrow on the right once more to reduce the size of the worksheet to 90 % of its normal size.

Notice that the "dotted lines" on your computer screen in Fig. 2.12 are now below Jennifer's name to indicate that all of the information, including her name, is now formatted to fit onto just one page when you print this file.

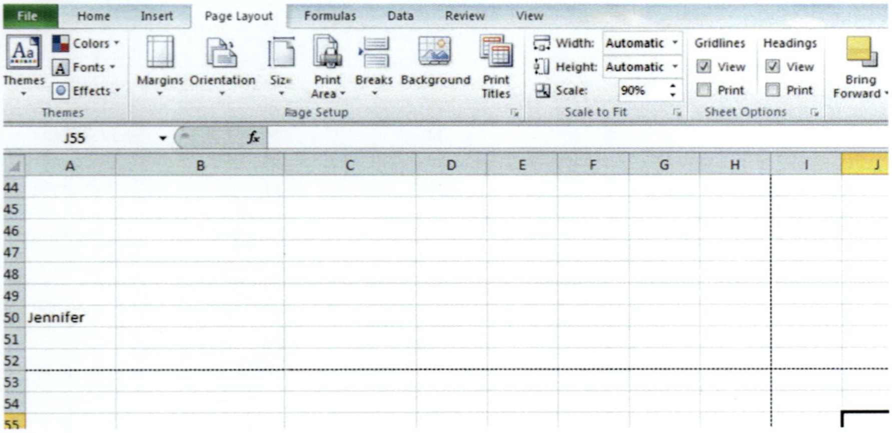

Fig. 2.12 Example of Scale Reduced to 90 % with "Jennifer" to be printed on the first page (note the *dotted line* below Jennifer on your screen)

Save the file as: Random34

Print the file. Does it all fit onto one page? It should (see Fig. 2.13).

Fig. 2.13 Final Spreadsheet of 90 % Scale to Fit

FRAME NO.	DUPLICATE FRAME NO.	RANDOM NO.
1	5	0.747176905
2	6	0.038774393
3	15	0.091368861
4	16	0.63147137
5	32	0.190734495
6	1	0.411943765
7	26	0.138033007
8	28	0.927874602
9	24	0.058336576
10	11	0.043243606
11	2	0.729011126
12	25	0.204119693
13	8	0.456656709
14	7	0.232589896
15	9	0.09096704
16	22	0.935399501
17	10	0.201267198
18	3	0.52638312
19	19	0.53734605
20	30	0.969840616
21	29	0.475657455
22	27	0.558049277
23	17	0.488444809
24	21	0.717097206
25	23	0.86192944
26	13	0.875595013
27	20	0.536748908
28	12	0.331784725
29	4	0.642847666
30	14	0.575767804
31	18	0.939789757
32	31	0.776050794
Jennifer		

2.5 End-of-Chapter Practice Problems

1. Suppose that you had cultured various strains of bacteria in a college microbiology laboratory. You want to study the growth rate of one strain of bacteria. You have samples of this bacteria strain in 63 test tubes and you would like to test randomly 15 of these 63 samples.

   (a) Set up a spreadsheet of frame numbers for these test tubes with the heading: FRAME NUMBERS using the Home/Fill commands.
   (b) Then, create a separate column to the right of these frame numbers which duplicates these frame numbers with the title: Duplicate frame numbers
   (c) Then, create a separate column to the right of these duplicate frame numbers and use the =RAND() function to assign random numbers to all of the frame numbers in the duplicate frame numbers column, and change this column format so that three decimal places appear for each random number
   (d) Sort the duplicate frame numbers and random numbers into a random order
   (e) Print the result so that the spreadsheet fits onto one page
   (f) Circle on your printout the I.D. number of the first 15 test tubes that you would use in your research study
   (g) Save the file as: RAND9

   Important note: Note that everyone who does this problem will generate a different random order of test tube ID numbers since Excel assigns a different random number each time the RAND() command is used. For this reason, the answer to this problem given in this Excel Guide will have a completely different sequence of random numbers from the random sequence that you generate. This is normal and what is to be expected.

2. Suppose that you have been hired as a consultant to test building materials for engineers designing suspension bridges. The engineers of this company are interested in using a new type of steel in future bridge construction. You have been given 114 samples of this type of steel and you have been asked to test a random sample of ten of these samples for tensile strength in terms of their material consistency.

   (a) Set up a spreadsheet of frame numbers for these steel samples with the heading: FRAME NO.
   (b) Then, create a separate column to the right of these frame numbers which duplicates these frame numbers with the title: Duplicate frame no.
   (c) Then, create a separate column to the right of these duplicate frame numbers entitled "Random number" and use the =RAND() function to assign random numbers to all of the frame numbers in the duplicate frame numbers column. Then, change this column format so that three decimal places appear for each random number

(d) Sort the duplicate frame numbers and random numbers into a random order
(e) Print the result so that the spreadsheet fits onto one page
(f) Circle on your printout the I.D. number of the first ten steel samples that would be used in this research study.
(g) Save the file as: RANDOM6

3. Suppose that a biology field researcher wants to take a random sample of 20 of 75 wild field mice that have been collected from the prairie grass that grows above the bluffs along the Mississippi River in Elsah, Illinois for a field research study.

(a) Set up a spreadsheet of frame numbers for these mice with the heading: FRAME NUMBERS.
(b) Then, create a separate column to the right of these frame numbers which duplicates these frame numbers with the title: Duplicate frame numbers
(c) Then, create a separate column to the right of these duplicate frame numbers entitled "Random number" and use the =RAND() function to assign random numbers to all of the frame numbers in the duplicate frame numbers column. Then, change this column format so that three decimal places appear for each random number
(d) Sort the duplicate frame numbers and random numbers into a random order
(e) Print the result so that the spreadsheet fits onto one page
(f) Circle on your printout the I.D. number of the first 20 mice that the field biologist should select for her study.
(g) Save the file as: RAND5

Chapter 3
Confidence Interval About the Mean Using the TINV Function and Hypothesis Testing

This chapter focuses on two ideas: (1) finding the 95 % confidence interval about the mean, and (2) hypothesis testing.

Let's talk about the confidence interval first.

3.1 Confidence Interval About the Mean

In statistics, we are always interested in *estimating the population mean*. How do we do that?

3.1.1 How to Estimate the Population Mean

Objective: To estimate the population mean, μ

Remember that the population mean is the average of all of the people in the target population. For example, if we were interested in how well adults ages 25–44 liked a new flavor of Ben & Jerry's ice cream, we could never ask this question of all of the people in the U.S. who were in that age group. Such a research study would take way too much time to complete and the cost of doing that study would be prohibitive.

So, instead of testing *everyone* in the population, we take a sample of people in the population and use the results of this sample to estimate the mean of the entire population. This saves both time and money. When we use the results of a sample to estimate the population mean, this is called "*inferential statistics*" because we are inferring the population mean from the sample mean.

T.J. Quirk et al., *Excel 2016 for Biological and Life Sciences Statistics*, Excel for Statistics, DOI 10.1007/978-3-319-39489-3_3

When we study a sample of people in science research, we know the size of our sample (n), the mean of our sample ($\overline{X}$), and the standard deviation of our sample (STDEV). We use these figures to estimate the population mean with a test called the "confidence interval about the mean."

3.1.2 Estimating the Lower Limit and the Upper Limit of the 95 % Confidence Interval About the Mean

The theoretical background of this test is beyond the scope of this book, and you can learn more about this test from studying any good statistics textbook (e.g. Levine 2011 or Bremer and Doerge 2010), but the basic ideas are as follows.

We assume that the population mean is somewhere in an interval which has a "lower limit" and an "upper limit" to it. We also assume in this book that we want to be "95 % confident" that the population mean is inside this interval somewhere. So, we intend to make the following type of statement:

> "We are 95 % confident that the population mean in miles per gallon (mpg) for the Chevy Impala automobile is between 26.92 and 29.42 miles per gallon."

If we want to create a billboard emphasizing the perceived lower environmental impact of the Chevy Impala by claiming that this car gets 28 miles per gallon (mpg), we can do this because 28 is *inside the 95 % confidence interval* in our research study in the above example. We do not know exactly what the population mean is, only that it is somewhere between 26.92 and 29.42 mpg, and 28 is inside this interval.

But we are only 95 % confident that the population mean is inside this interval, and 5 % of the time we will be wrong in assuming that the population mean is 28 mpg.

But, for our purposes in science research, we are happy to be 95 % confident that our assumption is accurate. We should also point out that 95 % is an arbitrary level of confidence for our results. We could choose to be 80 % confident, or 90 % confident, or even 99 % confident in our results if we wanted to do that. But, in this book, *we will always assume that we want to be 95 % confident of our results.* That way, you will not have to guess on how confident you want to be in any of the problems in this book. We will always want to be 95 % confident of our results in this book.

So how do we find the 95 % confidence interval about the mean for our data? In words, we will find this interval this way:

> "Take the sample mean ($\overline{X}$), *and add to it* 1.96 times the standard error of the mean (s.e.) to get the upper limit of the confidence interval. Then, take the sample mean, *and subtract from it* 1.96 times the standard error of the mean to get the lower limit of the confidence interval."

You will remember (See Sect. 1.3) that the standard error of the mean (s.e.) is found by dividing the standard deviation of our sample (STDEV) by the square root of our sample size, n.

In mathematical terms, the formula for the 95 % confidence interval about the mean is:

$$\overline{X} \pm 1.96 \text{ s.e.} \tag{3.1}$$

Note that the "± *sign*" stands for "plus or minus," and this means that you first add 1.96 times the s.e. to the mean to get the upper limit of the confidence interval, and then subtract 1.96 times the s.e. from the mean to get the lower limit of the confidence interval. Also, the symbol 1.96 s.e. means that you multiply 1.96 times the standard error of the mean to get this part of the formula for the confidence interval.

Note: We will explain shortly where the number 1.96 came from

Let's try a simple example to illustrate this formula.

3.1.3 Estimating the Confidence Interval the Chevy Impala in Miles Per Gallon

Let's suppose that you have been asked to be a part of a larger study looking at the carbon footprint of Chevy Impala drivers. You are interested in the average miles per gallon (mpg) of a Chevy Impala. You asked owners of the Chevy Impala to keep track of their mileage and the number of gallons used for two tanks of gas. Let's suppose that 49 owners did this, and that they average 27.83 miles per gallon (mpg) with a standard deviation of 3.01 mpg. The standard error (s.e.) would be 3.01 divided by the square root of 49 (i.e., 7) which gives a s.e. equal to 0.43.

The 95 % confidence interval for these data would be:

$$27.83 \pm 1.96\ (0.43)$$

The *upper limit of this confidence interval* uses the plus sign of the ± sign in the formula. Therefore, the upper limit would be:

$$27.83 + 1.96\ (0.43) = 27.83 + 0.84 = 28.67 \text{ mpg}$$

Similarly, *the lower limit of this confidence interval* uses the minus sign of the ± sign in the formula. Therefore, the lower limit would be:

$$27.83 - 1.96\ (0.43) = 27.83 - 0.84 = 26.99 \text{ mpg}$$

The result of our part of the ongoing research study would, therefore, be the following:

> "We are 95 % confident that the population mean for the Chevy Impala is somewhere between 26.99 mpg and 28.67 mpg."

We could create a billboard emphasizing 28 miles per gallon and highlight a perceived lower environmental impact. Our data supports this claim because the 28 mpg is inside of this 95 % confidence interval for the population mean.

You are probably asking yourself: "Where did that 1.96 in the formula come from?"

3.1.4 Where Did the Number "1.96" Come From?

A detailed mathematical answer to that question is beyond the scope of this book, but here is the basic idea.

We make an assumption that the data in the population are "normally distributed" in the sense that the population data would take the shape of a "normal curve" if we could test all of the people in the population. The normal curve looks like the outline of the Liberty Bell that sits in front of Independence Hall in Philadelphia, Pennsylvania. The normal curve is "symmetric" in the sense that if we cut it down the middle, and folded it over to one side, the half that we folded over would fit perfectly onto the half on the other side.

A discussion of integral calculus is beyond the scope of this book, but essentially we want to find the lower limit and the upper limit of the population data in the normal curve so that 95 % of the area under this curve is between these two limits. *If we have more than 40 people in our research study*, the value of these limits is plus or minus 1.96 times the standard error of the mean (s.e.) of our sample. The number 1.96 times the s.e. of our sample gives us the upper limit and the lower limit of our confidence interval. If you want to learn more about this idea, you can consult a good statistics book (e.g. Salkind 2010 or van Emden 2008).

The number 1.96 would change if we wanted to be confident of our results at a different level from 95 % as long as we have more than 40 people in our research study.

For example:

1. If we wanted to be 80 % confident of our results, this number would be 1.282.
2. If we wanted to be 90 % confident of our results, this number would be 1.645.
3. If we wanted to be 99 % confident of our results, this number would be 2.576.

But since we always want to be 95 % confident of our results in this book, we will always use 1.96 in this book whenever we have more than 40 people in our research study.

By now, you are probably asking yourself: "Is this number in the confidence interval about the mean always 1.96?" The answer is: "No!", and we will explain why this is true now.

3.1.5 *Finding the Value for t in the Confidence Interval Formula*

Objective: To find the value for t in the confidence interval formula

The correct formula for the confidence interval about the mean for different sample sizes is the following:

$$\overline{X} \pm \text{t s.e.} \qquad (3.2)$$

To use this formula, you find the sample mean, $\overline{X}$, *and add to it the value of t times the s.e. to get the upper limit* of this 95 % confidence interval. Also, you take the sample mean, $\overline{X}$, and *subtract from it the value of t times the s.e. to get the lower limit* of this 95 % confidence interval. And, you find the value of t in the table given in Appendix E of this book in the following way:

Objective: To find the value of t in the t-table in Appendix E

Before we get into an explanation of what is meant by "the value of t," let's give you practice in finding the value of t by using the t-table in Appendix E.

Keep your finger on Appendix E as we explain how you need to "read" that table.

Since the test in this chapter is called the "confidence interval about the mean test," you will use the first column on the left in Appendix E to find the critical value of t for your research study (note that this column is headed: "sample size n").

To find the value of *t*, you go down this first column until you find the sample size in your research study, and then you go to the right and read the value of *t* for that sample size in the "critical t column" of the table (note that this column is the column that you would use for the 95 % confidence interval about the mean).

For example, if you have 14 people in your research study, the value of t is 2.160.

If you have 26 people in your research study, the value of t is 2.060.

If you have more than 40 people in your research study, the value of t is always 1.96.

Note that the "critical t column" in Appendix E represents the value of t that you need to use to obtain to be 95 % confident of your results as "significant" results.

Throughout this book, we are assuming that you want to be 95 % confident in the results of your statistical tests. Therefore, the value for t in the t-table in Appendix E tells you which value you should use for t when you use the formula for the 95 % confidence interval about the mean.

Now that you know how to find the value of t in the formula for the confidence interval about the mean, let's explore how you find this confidence interval using Excel.

3.1.6 Using Excel's TINV Function to Find the Confidence Interval About the Mean

Objective: To use the TINV function in Excel to find the confidence interval about the mean

When you use Excel, the formulas for finding the confidence interval are:

$$\textit{Lower limit}: \; = \overline{X} - \textit{TINV}(1 - 0.95, n - 1)^{*}s.e. \quad \text{(no spaces between these symbols)} \tag{3.3}$$

$$\textit{Upper limit}: \; = \overline{X} + \textit{TINV}(1 - 0.95, n - 1)^{*}s.e. \quad \text{(no spaces between these symbols)} \tag{3.4}$$

Note that the "* *symbol*" in this formula tells Excel to use the multiplication step in the formula, and it stands for "times" in the way we talk about multiplication.

You will recall from Chap. 1 that n stands for the sample size, and so $n-1$ stands for the sample size minus one.

You will also recall from Chap. 1 that the standard error of the mean, s.e., equals the STDEV divided by the square root of the sample size, n (see Sect. 1.3).

Let's try a sample problem using Excel to find the 95 % confidence interval about the mean for a problem.

Let's suppose that General Motors wanted to claim that its Chevy Impala achieves 28 miles per gallon (mpg). Let's call 28 mpg the "reference value" for this car.

Suppose that you work for Ford Motor Co. and that you want to check this claim to see is it holds up based on some research evidence. You decide to collect some data and to use a two-side 95 % confidence interval about the mean to test your results:

3.1.7 *Using Excel to find the 95 % Confidence Interval for a Car's mpg Claim*

Objective: To analyze the data using a two-side 95 % confidence interval about the mean

You select a sample of new car owners for this car and they agree to keep track of their mileage for two tanks of gas and to record the average miles per gallon they achieve on these two tanks of gas. Your research study produces the results given in Fig. 3.1:

Chevy Impala
Miles per gallon
30.9
24.5
31.2
28.7
35.1
29.0
28.8
23.1
31.0
30.2
28.4
29.3
24.2
27.0
26.7
31.0
23.5
29.4
26.3
27.5
28.2
28.4
29.1
21.9
30.9

Fig. 3.1 Worksheet Data for Chevy Impala (Practical Example)

Create a spreadsheet with these data and use Excel to find the sample size (n), the mean, the standard deviation (STDEV), and the standard error of the mean (s.e.) for these data using the following cell references.

A3: Chevy Impala
A5: Miles per gallon
A6: 30.9

Enter the other mpg data in cells A7:A30

Now, highlight cells A6:A30 and format these numbers in number format (one decimal place). Center these numbers in Column A. Then, widen columns A and B by making both of them twice as wide as the original width of column A. Then, widen column C so that it is three times as wide as the original width of column A so that your table looks more professional.

C7: n
C10: Mean
C13: STDEV
C16: s.e.
C19: 95 % confidence interval
D21: Lower limit:
D23: Upper limit: (see Fig. 3.2)

Chevy Impala					
Miles per gallon					
30.9					
24.5		**n**			
31.2					
28.7					
35.1		**Mean**			
29.0					
28.8					
23.1		**STDEV**			
31.0					
30.2					
28.4		**s.e**			
29.3					
24.2					
27.0		**95% confidence interval**			
26.7					
31.0			**Lower limit:**		
23.5					
29.4			**Upper Limit:**		
26.3					
27.5					
28.2					
28.4					
29.1					
21.9					
30.9					

Fig. 3.2 Example of Chevy Impala Format for the Confidence Interval About the Mean Labels

B26: Draw a picture below this confidence interval
B28: 26.92
B29: lower (right-align this word)
B30: limit (right-align this word)
C28: '-------- 28 -------–28.17 ----------– (note that you need to begin cell C28 with a *single quotation mark* (') to tell Excel that this is a *label*, and not a number)
D28: '----------- (note the single quotation mark)
E28: '29.42 (note the single quotation mark)
C29: ref. Mean
C30: value
E29: upper
E30: limit
B33: Conclusion:

Now, align the labels underneath the picture of the confidence interval so that they look like Fig. 3.3.

Chevy Impala					
Miles per gallon					
30.9					
24.5		**n**			
31.2					
28.7					
35.1		**Mean**			
29.0					
28.8					
23.1		**STDEV**			
31.0					
30.2					
28.4		**s.e**			
29.3					
24.2					
27.0		**95% confidence interval**			
26.7					
31.0			**Lower limit:**		
23.5					
29.4			**Upper Limit:**		
26.3					
27.5					
28.2	**Draw a picture below this confidence interval**				
28.4					
29.1	**26.92 ---------**	**28 --------- 28.17 ---------------**		**29.42**	
21.9	**lower**	**ref. Mean**		**upper**	
30.9	**limit**	**value**		**limit**	
	Conclusion:				

Fig. 3.3 Example of Drawing a Picture of a Confidence Interval About the Mean Result

Next, name the range of data from A6:A30 as: miles

D7: Use Excel to find the sample size
D10: Use Excel to find the mean
D13: Use Excel to find the STDEV
D16: Use Excel to find the s.e.

Now, you need to find the lower limit and the upper limit of the 95 % confidence interval for this study.

We will use Excel's TINV function to do this. We will assume that you want to be 95 % confident of your results.

F21 : =D10−TINV(1−.95,24)*D16 (do not use spaces between these symbols)

Note that this TINV formula uses 24 since 24 is one less than the sample size of 25 (i.e., 24 is n−1). Note that D10 is the mean, while D16 is the standard error of the mean. The above formula gives the *lower limit of the confidence interval, 26.92.*

F23 : =D10+TINV(1−.95,24)*D16 (do not use spaces between these symbols)

The above formula gives the *upper limit of the confidence interval, 29.42.*

Now, use number format (two decimal places) in your Excel spreadsheet for the mean, standard deviation, standard error of the mean, and for both the lower limit and the upper limit of your confidence interval. If you printed this spreadsheet now, the lower limit of the confidence interval (26.92) and the upper limit of the confidence interval (29.42) would "dribble over" onto a second printed page because the information on the spreadsheet is too large to fit onto one page in its present format.

So, you need to use Excel's "Scale to Fit" commands that we discussed in Chap. 2 (see Sect. 2.4) to reduce the size of the spreadsheet to 95 % of its current size using the Page Layout/Scale to Fit function. Do that now, and notice that the dotted line to the right of 26.92 and 29.42 indicates that these numbers would now fit onto one page when the spreadsheet is printed out (see Fig. 3.4)

Chevy Impala					
Miles per gallon					
30.9					
24.5		**n**	**25**		
31.2					
28.7					
35.1		**Mean**	**28.17**		
29.0					
28.8					
23.1		**STDEV**	**3.03**		
31.0					
30.2					
28.4		**s.e**	**0.61**		
29.3					
24.2					
27.0		**95% confidence interval**			
26.7					
31.0			**Lower limit:**		**26.92**
23.5					
29.4			**Upper Limit:**		**29.42**
26.3					
27.5					
28.2	**Draw a picture below this confidence interval**				
28.4					
29.1	**26.92**	**---------- 28 ---------- 28.17 ----------------**		**29.42**	
21.9	**lower**	**ref. Mean**		**upper**	
30.9	**limit**	**value**		**limit**	
	Conclusion:				

Fig. 3.4 Result of Using the TINV Function to Find the Confidence Interval About the Mean

Note that you have drawn a picture of the 95 % confidence interval beneath cell B26, including the lower limit, the upper limit, the mean, and the reference value of 28 mpg given in the claim that the company wants to make about the car's miles per gallon performance.

Now, let's write the conclusion to your research study on your spreadsheet:

C33: Since the reference value of 28 is inside
C34: the confidence interval, we accept that
C35: the Chevy Impala does get 28 mpg.

Important note: *You are probably wondering why we wrote the conclusion on three separate lines of the spreadsheet instead of writing it on one long line. This is because if you wrote it on one line, two things would happen that you would not like: (1) If you printed the conclusion by reducing the size of the layout of the page so that the entire spreadsheet would fit onto one page, the print font size for the entire spreadsheet would be so small that you could not read it without a magnifying glass, and (2) If you printed the spreadsheet without reducing the page size layout, it would "dribble over" part of the conclusion to a separate page all by itself, and your spreadsheet would not look professional.*

Your research study accepted the claim that the Chevy Impala did get 28 miles per gallon. The average miles per gallon in your study was 28.17. (See Fig. 3.5)

Save your resulting spreadsheet as: **CHEVY7**

Chevy Impala					
Miles per gallon					
30.9					
24.5		**n**	**25**		
31.2					
28.7					
35.1		**Mean**	**28.17**		
29.0					
28.8					
23.1		**STDEV**	**3.03**		
31.0					
30.2					
28.4		**s.e**	**0.61**		
29.3					
24.2					
27.0		**95% confidence interval**			
26.7					
31.0			**Lower limit:**		**26.92**
23.5					
29.4			**Upper Limit:**		**29.42**
26.3					
27.5					
28.2	**Draw a picture below this confidence interval**				
28.4					
29.1	**26.92 ----------**	**28 ---------- 28.17 ---------------**		**29.42**	
21.9	**lower**	**ref. Mean**		**upper**	
30.9	**limit**	**value**		**limit**	
	Conclusion:	**Since the reference value of 28 is inside**			
		the confidence interval, we accept that			
		the Chevy Impala does get 28 mpg.			

Fig. 3.5 Final Spreadsheet for the Chevy Impala Confidence Interval About the Mean

3.2 Hypothesis Testing

One of the important activities of research scientists is that they attempt to "check" their assumptions about the world by testing these assumptions in the form of hypotheses.

A typical hypothesis is in the form: *"If x, then y."*

Some examples would be:

1. "If we use this new method fertilizing the soil, the corn yield of the plot will increase by 3 %."
2. "If we use this new method of teaching science to ninth graders, then our science achievement scores will go up by 5 %."
3. "If we change the format for teaching Introductory Biology to our undergraduates, then their final exam scores will increase by 8 %."

A hypothesis, then, to a research scientist is a "guess" about what we think is true in the real world. We can test these guesses using statistical formulas to see if our predictions come true in the real world.

So, in order to perform these statistical tests, we must first state our hypotheses so that we can test our results against our hypotheses to see if our hypotheses match reality.

So, how do we generate hypotheses in science research?

3.2.1 Hypotheses Always Refer to the Population of People, Plants, or Animals that You Are Studying

The first step is to understand that our hypotheses always refer to the *population* of people, plants, or animals under study.

For example, if we are interested in studying a species of noxious weed found along highways of southern South Dakota, we would select various sections of highways and estimate the number of weeds found in these sections, these sections would be used as our sample. This sample would be used in generalizing our findings for all of the highways in southern South Dakota.

All of the highways in southern south Dakota would be the *population* that we are interested in studying, while the particular sections of highways in our study are called the *sample* from this population.

Since our sample sizes typically contain only a portion of the highways, we are interested in the results of our sample *only insofar as the results of our sample can be "generalized" to the population in which we are really interested.*

That is why our hypotheses always refer to the population, and never to the sample of people, plants, animals, or events in our study.

You will recall from Chap. 1 that we used the symbol: $\overline{X}$ to refer to the mean of the sample we use in our research study (See Sect. 1.1).

We will use the symbol: μ (the Greek letter "mu") to refer to the *population mean.*

In testing our hypotheses, we are trying to decide which one of two competing hypotheses *about the population mean* we should accept given our data set.

3.2.2 The Null Hypothesis and the Research (Alternative) Hypothesis

These two hypotheses are called the *null hypothesis* and the *research hypothesis.*

Statistics textbooks typically refer to the *null hypothesis* with the notation: H_0.

The *research hypothesis* is typically referred to with the notation: H_1 and it is sometimes called the *alternative hypothesis.*

Let's explain first what is meant by the null hypothesis and the research hypothesis:

(1) *The null hypothesis is what we accept as true unless we have compelling evidence that it is not true.*
(2) *The research hypothesis is what we accept as true whenever we reject the null hypothesis as true.*

This is similar to our legal system in America where we assume that a supposed criminal is innocent until he or she is proven guilty in the eyes of a jury. Our null hypothesis is that this defendant is innocent, while the research hypothesis is that he or she is guilty.

In the great state of Missouri, every license plate has the state slogan: "Show me." This means that people in Missouri think of themselves as not gullible enough to accept everything that someone says as true unless that person's actions indicate the truth of his or her claim. In other words, people in Missouri believe strongly that a person's actions speak much louder than that person's words.

Since both the null hypothesis and the research hypothesis cannot both be true, the task of hypothesis testing using statistical formulas is to decide which one you will accept as true, and which one you will reject as true.

Sometimes in science research a series of rating scales is used to measure people's attitudes toward a company, toward one of its products, or toward their intention-to-buy that company's products. These rating scales are typically 5-point, 7-point, or 10-point scales, although other scale values are often used as well.

3.2.2.1 Determining the Null Hypothesis and the Research Hypothesis When Rating Scales Are Used

Here is a typical example of a 7-point scale in science education for parents of 8th grade pupils at the end of a school year (see Fig. 3.6):

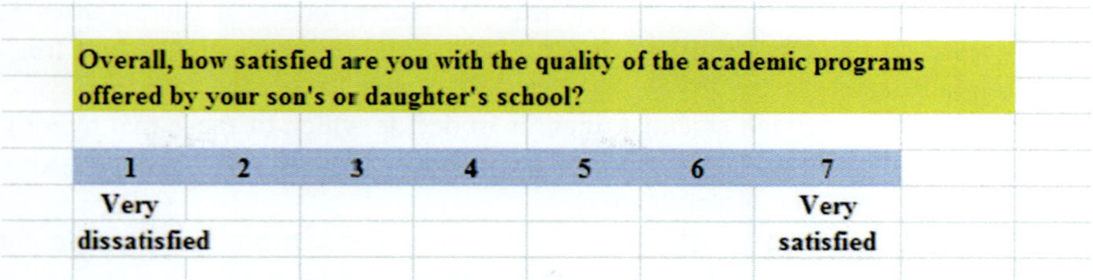

Fig. 3.6 Example of a Rating Scale Item for Parents of 8th Graders (Practical Example)

So, how do we decide what to use as the null hypothesis and the research hypothesis whenever rating scales are used?

> Objective: To decide on the null hypothesis and the research hypothesis whenever rating scales are used.

In order to make this determination, we will use a simple rule:

Rule: Whenever rating scales are used, we will use the "middle" of the scale as the null hypothesis and the research hypothesis.

In the above example, since 4 is the number in the middle of the scale (i.e., three numbers are below it, and three numbers are above it), our hypotheses become:

Null hypothesis: $\mu = 4$
Research hypothesis: $\mu \neq 4$

In the above rating scale example, if the result of our statistical test for this one attitude scale item indicates that our population mean is "close to 4," we say that we accept the null hypothesis that the parents of 8th grade pupils were neither satisfied nor dissatisfied with the quality of the science program offered by their son's or daughter's school.

In the above example, *if the result of our statistical test indicates that the population mean is significantly different from 4*, we reject the null hypothesis and accept the research hypothesis *by stating either that*:

"*Parents of 8th grade pupils were significantly satisfied with the quality of the science program offered by their son's or daughter's school*" (this is true whenever our sample mean is significantly greater than our expected population mean of 4).

or

"*Parents of 8th grade pupils were significantly dissatisfied with the quality of the science program offered by their son's or daughter's school*" (this is accepted as true whenever our sample mean is significantly less than our expected population mean of 4).

Both of these conclusions cannot be true. We accept one of the hypotheses as "true" based on the data set in our research study, and the other one as "not true" based on our data set.

The job of the research scientist, then, is to decide which of these two hypotheses, the null hypothesis or the research hypothesis, he or she will accept as true given the data set in the research study.

Let's try some examples of rating scales so that you can practice figuring out what the null hypothesis and the research hypothesis are for each rating scale.

In the spaces in Fig. 3.7, write in the null hypothesis and the research hypothesis for the rating scales:

1. State University is an excellent university.

1	2	3	4	5
Strongly Disagree	Disagree	Undecided	Agree	Strongly Agree

Null hypothesis: $\mu =$ ______

Research hypothesis: $\mu \neq$ ______

2. How would you rate the quality of teaching in the Environmental Science department at State University?

poor	1	2	3	4	5	6	7	excellent

Null hypothesis: $\mu =$ ______

Research hypothesis: $\mu \neq$ ______

3. How would you rate the quality of faculty in the Environmental Science department at State University?

1	2	3	4	5	6	7	8	9	10
very poor									very good

Null hypothesis: $\mu =$ ______

Research hypothesis: $\mu \neq$ ______

Fig. 3.7 Examples of Rating Scales for Determining the Null Hypothesis and the Research Hypothesis

How did you do?

Here are the answers to these three questions:

1. The null hypothesis is $\mu = 3$, and the research hypothesis is $\mu \neq 3$ on this 5-point scale (i.e. the "middle" of the scale is 3).

2. The null hypothesis is $\mu = 4$, and the research hypothesis is $\mu \neq 4$ on this 7-point scale (i.e., the "middle" of the scale is 4).
3. The null hypothesis is $\mu = 5.5$, and the research hypothesis is $\mu \neq 5.5$ on this 10-point scale (i.e., the "middle" of the scale is 5.5 since there are 5 numbers below 5.5 and 5 numbers above 5.5).

As another example, suppose Texas Parks and Wildlife uses a 4-point scale in its post-hunting satisfaction survey. The results of this survey are used to determine the number of licenses issued for wildlife management the following hunting season. The scale is as follows:

1 = Not So Good
2 = Average
3 = Very Good
4 = Great

On this scale, the null hypothesis is: $\mu = 2.5$ and the research hypothesis is: $\mu \neq 2.5$, because there are two numbers below 2.5, and two numbers above 2.5 on that rating scale.

Now, let's discuss the 7 STEPS of hypothesis testing for using the confidence interval about the mean.

3.2.3 *The 7 Steps for Hypothesis-Testing Using the Confidence Interval About the Mean*

Objective:	To learn the 7 steps of hypothesis-testing using the confidence interval about the mean

There are seven basic steps of hypothesis-testing for this statistical test.

3.2.3.1 STEP 1: State the Null Hypothesis and the Research Hypothesis

If you are using numerical scales in your survey, you need to remember that these hypotheses refer to the "middle" of the numerical scale. For example, if you are using 7-point scales with 1 = poor and 7 = excellent, these hypotheses would refer to the middle of these scales and would be:

$$\text{Null hypothesis } H_0: \mu = 4$$
$$\text{Research hypothesis } H_1: \mu \neq 4$$

3.2.3.2 STEP 2: Select the Appropriate Statistical Test

In this chapter we are studying the confidence interval about the mean, and so we will select that test.

3.2.3.3 STEP 3: Calculate the Formula for the Statistical Test

You will recall (see Sect. 3.1.5) that the formula for the confidence interval about the mean is:

$$\overline{X} \pm \text{t s.e.} \tag{3.2}$$

We discussed the procedure for computing this formula for the confidence interval about the mean using Excel earlier in this chapter, and the steps involved in using that formula are:

1. Use Excel's = COUNT function to find the sample size.
2. Use Excel's = AVERAGE function to find the sample mean, $\overline{X}$.
3. Use Excel's = STDEV function to find the standard deviation, STDEV.
4. Find the standard error of the mean (s.e.) by dividing the standard deviation (STDEV) by the square root of the sample size, n.
5. Use Excel's TINV function to find the lower limit of the confidence interval.
6. Use Excel's TINV function to find the upper limit of the confidence interval.

3.2.3.4 STEP 4: Draw a Picture of the Confidence Interval About the Mean, Including the Mean, the Lower Limit of the Interval, the Upper Limit of the Interval, and the Reference Value Given in the Null Hypothesis, H_0 (We Will Explain Step 4 Later in the Chapter.)

3.2.3.5 STEP 5: Decide on a Decision Rule

(a) *If the reference value is inside the confidence interval, accept the null hypothesis, H_0*
(b) *If the reference value is outside the confidence interval, reject the null hypothesis, H_0, and accept the research hypothesis, H_1*

3.2.3.6 STEP 6: State the Result of your Statistical Test

There are two possible results when you use the confidence interval about the mean, and only one of them can be accepted as "true." So your result would be one of the following:

Either: Since the reference value is inside the confidence interval, *we accept the null hypothesis,* H_0

Or: Since the reference value is outside the confidence interval, *we reject the null hypothesis,* H_0*, and accept the research hypothesis,* H_1

3.2.3.7 STEP 7: State the Conclusion of Your Statistical Test in Plain English!

In practice, this is more difficult than it sounds because you are trying to summarize the result of your statistical test in simple English that is both concise and accurate so that someone who has never had a statistics course (such as your boss, perhaps) can understand the conclusion of your test. This is a difficult task, and we will give you lots of practice doing this last and most important step throughout this book.

Objective: To write the conclusion of the confidence interval about the mean test

Let's set some basic rules for stating the conclusion of a hypothesis test.

Rule #1: Whenever you reject H_0 *and accept* H_1*, you must use the word "significantly" in the conclusion to alert the reader that this test found an important result.*

Rule #2: Create an outline in words of the "key terms" you want to include in your conclusion so that you do not forget to include some of them.

Rule #3: Write the conclusion in plain English so that the reader can understand it even if that reader has never taken a statistics course.

Let's practice these rules using the Chevy Impala Excel spreadsheet that you created earlier in this chapter, but first we need to state the hypotheses for that car.

If General Motors wants to claim that the Chevy Impala gets 28 miles per gallon on a billboard ad, the hypotheses would be:

H_0: $\mu = 28$ mpg
H_1: $\mu \neq 28$ mpg

You will remember that the reference value of 28 mpg was inside the 95 % confidence interval about the mean for your data, so we would accept H_0 for the Chevy Impala that the car does get 28 mpg.

Objective: To state the result when you accept H_0

Result: Since the reference value of 28 mpg is inside the confidence interval, we accept the null hypothesis, H_0

Let's try our three rules now:

Objective: To write the conclusion when you accept H_0

Rule #1*: Since the reference value was inside the confidence interval, we cannot use the word "significantly" in the conclusion. This is a basic rule we are using in this chapter for every problem.*

Rule #2: The key terms in the conclusion would be:

- Chevy Impala
- reference value of 28 mpg

Rule #3: The Chevy Impala did get 28 mpg.

The process of writing the conclusion when you accept H_0 is relatively straightforward since you put into words what you said when you wrote the null hypothesis.

However, the process of stating the conclusion when you reject H_0 and accept H_1 is more difficult, so let's practice writing that type of conclusion with three practice case examples:

Objective: To write the result and conclusion when you reject H_0

CASE #1: Suppose that an ad in *The Wall Street Journal* claimed that the Honda Accord Sedan gets 34 miles per gallon. The hypotheses would be:

H_0: $\mu = 34$ mpg
H_1: $\mu \neq 34$ mpg

Suppose that your research yields the following confidence interval:

30	31	32	34
lower limit	Mean	upper limit	Ref. Value

Result: Since the reference value is outside the confidence interval, we reject the null hypothesis and accept the research hypothesis

The three rules for stating the conclusion would be:

Rule #1: We must include the word "significantly" since the reference value of 34 is outside the confidence interval.

Rule #2: The key terms would be:

- Honda Accord Sedan
- significantly
- either "more than" or "less than"
- and probably closer to

Rule #3: The Honda Accord Sedan got significantly less than 34 mpg, and it was probably closer to 31 mpg.

Note that this conclusion says that the mpg was less than 34 mpg because the sample mean was only 31 mpg. Note, also, that when you find a significant result by rejecting the null hypothesis, *it is not sufficient to say only: "significantly less than 34 mpg,"* because that does not tell the reader "how much less than 34 mpg" the sample mean was from 34 mpg. To make the conclusion clear, you need to add: "probably closer to 31 mpg" since the sample mean was only 31 mpg.

CASE #2: The National Association of Biology Teachers (NABT) is an American-based scholarly Life Sciences society with more than 5,000 members. The NABT hosts an annual Professional Development Conference for 4 days that focuses on the teaching of biology and the life sciences in the twenty-first century. The conference includes hands-on workshops and informative sessions for participants. Suppose that the NABT wanted to use an Internet Survey to evaluate the annual conference based on responses from the participants. Let's suppose that you have been asked to perform the data analyses for the returned surveys, and that you want to practice your data analysis skills on the hypothetical Item #15 given in Fig. 3.8:

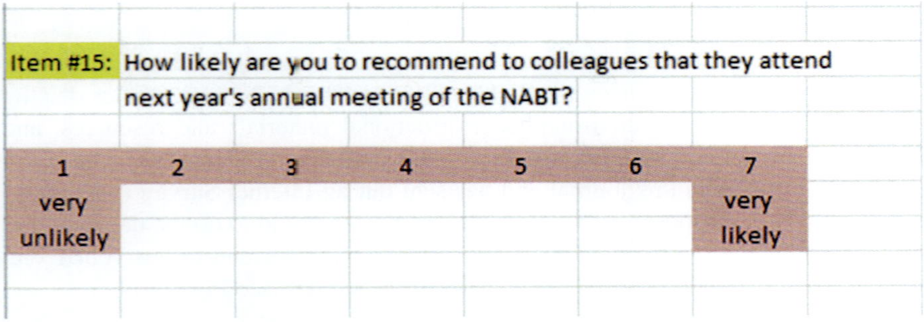

Fig. 3.8 Example of Item #15 of the NABT Survey

The hypotheses for this one item would be:

H_0: $\mu = 4$
H_1: $\mu \neq 4$

Essentially, the null hypothesis equal to 4 states that if the obtained mean score for this question is not significantly different from 4 on the rating scale, then attendees, overall, were neither likely nor unlikely to recommend to colleagues that they attend next year's annual conference.

Suppose that your analysis produced the following confidence interval for this item on the survey.

1.8	2.8	3.8	4
lower limit	Mean	upper limit	Ref. Value

Result: Since the reference value is outside the confidence interval, we reject the null hypothesis and accept the research hypothesis.

Rule #1: You must include the word "significantly" since the reference value is outside the confidence interval

Rule #2: The key terms would be:

- attendees
- significantly
- NABT annual meeting this year
- either likely or unlikely (since the result is significant)
- recommend to colleagues
- attend next year's annual meeting of the NABT

Rule #3: Attendees at this year's annual meeting of the NABT were significantly unlikely to recommend to colleagues that they attend next year's annual meeting of the NABT.

Note that you need to use the word "unlikely" since the sample mean of 2.8 was on the unlikely side of the middle of the rating scale.

CASE #3: The National Association of Biology Teachers (NABT) publishes the journal: *The American Biology Teacher* which includes articles dealing with ecology/environment, instructional materials and resources, and education. Suppose that the NABT wanted to evaluate the quality of articles in this journal and has sent out an Internet Survey to its members. Suppose, further, that you have been asked to do the data analysis for the returned surveys, and that you have decided to test your Excel skills on a hypothetical Item #8 of the survey (see Fig. 3.9)

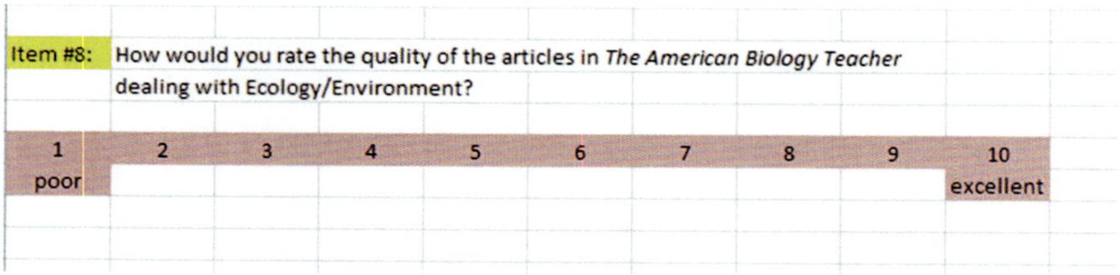

Fig. 3.9 Hypothetical Example of Item #8 of the NABT Internet Survey

This item would have the following hypotheses:

H_0: $\mu = 5.5$
H_1: $\mu \neq 5.5$

Suppose that your research produced the following confidence interval for this item on the survey:

5.5	5.7	5.8	5.9
Ref. Value	lower limit	Mean	upper limit

Result: Since the reference value is outside the confidence interval, we reject the null hypothesis and accept the research hypothesis

The three rules for stating the conclusion would be:

Rule #1: You must include the word "significantly" since the reference value is outside the confidence interval

Rule #2: The key terms would be:

- Members of the NABT
- *The American Biology Teacher*
- significantly
- rated the quality of articles
- Internet survey
- dealing with Ecology/Environment
- either "positive" or "negative" (we will explain this)

Rule #3: Members of the NABT rated the quality of articles in *The American Biology Teacher* dealing with Ecology/Environment in an Internet survey as significantly positive.

Note two important things about this conclusion above: (1) people when speaking English do not normally say "significantly excellent" since something is either excellent or is not excellent without any modifier, and (2) since the mean rating of the quality of the articles dealing with Ecology/Environment (5.8) was significantly greater than 5.5 on the positive side of the scale, we would say "significantly positive" to indicate this fact.

If you want a more detailed explanation of the confidence interval about the mean, see Townend (2002).

The three practice problems at the end of this chapter will give you additional practice in stating the conclusion of your result, and this book will include many more examples that will help you to write a clear and accurate conclusion to your research findings.

3.3 Alternative Ways to Summarize the Result of a Hypothesis Test

It is important for you to understand that in this book we are summarizing an hypothesis test in one of two ways: (1) We accept the null hypothesis, or (2) We reject the null hypothesis and accept the research hypothesis. We are consistent in the use of these words so that you can understand the concept underlying hypothesis testing.

However, there are many other ways to summarize the result of an hypothesis test, and all of them are correct theoretically, even though the terminology differs. If you are taking a course with a professor who wants you to summarize the results of a statistical test of hypotheses in language which is different from the language we are using in this book, do not panic! If you understand the concept of hypothesis testing as described in this book, you can then translate your understanding to use the terms that your professor wants you to use to reach the same conclusion to the hypothesis test.

Statisticians and professors of science statistics all have their own language that they like to use to summarize the results of an hypothesis test. There is no one set of words that these statisticians and professors will ever agree on, and so we have chosen the one that we believe to be easier to understand in terms of the concept of hypothesis testing.

To convince you that there are many ways to summarize the results of an hypothesis test, we present the following quotes from prominent statistics and research books to give you an idea of the different ways that are possible.

3.3.1 Different Ways to Accept the Null Hypothesis

The following quotes are typical of the language used in statistics and research books *when the null hypothesis is accepted*:

"The null hypothesis is not rejected." (Black 2010, p. 310)

"The null hypothesis cannot be rejected." (McDaniel and Gates 2010, p. 545)

"The null hypothesis ... claims that there is no difference between groups." (Salkind 2010, p. 193)

"The difference is not statistically significant." (McDaniel and Gates 2010, p. 545)

"... the obtained value is not extreme enough for us to say that the difference between Groups 1 and 2 occurred by anything other than chance." (Salkind 2010, p. 225)

"If we do not reject the null hypothesis, we conclude that there is not enough statistical evidence to infer that the alternative (hypothesis) is true." (Keller 2009, p. 358)

"The research hypothesis is not supported." (Zikmund and Babin 2010, p. 552)

3.3.2 Different Ways to Reject the Null Hypothesis

The following quotes are typical of the quotes used in statistics and research books *when the null hypothesis is rejected*:

"The null hypothesis is rejected." (McDaniel and Gates 2010, p. 546)

"If we reject the null hypothesis, we conclude that there is enough statistical evidence to infer that the alternative hypothesis is true." (Keller 2009, p. 358)

"If the test statistic's value is inconsistent with the null hypothesis, we reject the null hypothesis and infer that the alternative hypothesis is true." (Keller 2009, p. 348)

"Because the observed value ... is greater than the critical value ..., the decision is to reject the null hypothesis." (Black 2010, p. 359)

"If the obtained value is more extreme than the critical value, the null hypothesis cannot be accepted." (Salkind 2010, p. 243)

"The critical t-value ... must be surpassed by the observed t-value if the hypothesis test is to be statistically significant ...." (Zikmund and Babin 2010, p. 567)

"The calculated test statistic ... exceeds the upper boundary and falls into this rejection region. The null hypothesis is rejected." (Weiers 2011, p. 330)

You should note that all of the above quotes are used by statisticians and professors when discussing the results of an hypothesis test, and so you should not be surprised if someone asks you to summarize the results of a statistical test using a different language than the one we are using in this book.

3.4 End-of-Chapter Practice Problems

1. Suppose that you were hired to conduct a research study determining milk yield in bovines on a dairy farm in the state of Wisconsin. Last year, the average annual milk yield per cow on this farm was 6,800 pounds (lbs.) in 2-year-old registered Jersey cows. You have been asked to take a random sample of cows and to check their annual milk yield at the end of the current year. You decide to test your Excel skills on a small number of cows, and the hypothetical data (in hundreds of pounds) appear in Fig. 3.10:

Fig. 3.10 Worksheet Data for Chap. 3: Practice Problem #1

MILK YIELD OF JERSEY COWS IN WISCONSIN

ANNUAL YIELD (in hundreds of pounds)
100
90
78
70
56
52
50
55
58
62
63
56
54
52
53
51
56
50
51

(a) To the right of this table, use Excel to find the sample size, mean, standard deviation, and standard error of the mean for the yield figures. Label your answers. Use number format (two decimal places) for the mean, standard deviation, and standard error of the mean.
(b) Enter the null hypothesis and the research hypothesis onto your spreadsheet.
(c) Use Excel's TINV function to find the 95 % confidence interval about the mean for these figures. Label your answers. Use number format (two decimal places).
(d) Enter your *result* onto your spreadsheet.
(e) Enter your *conclusion in plain English* onto your spreadsheet.
(f) Print the final spreadsheet to fit onto one page (if you need help remembering how to do this, see the objectives at the end of Chap. 2 in Sect. 2.4)
(g) On your printout, draw a diagram of this 95 % confidence interval by hand
(h) Save the file as: yield3

2. Suppose that a fish hatchery in the state of Colorado has asked you to determine the average weight of the trout they are releasing into streams and lakes in Colorado. If the fish are too small, licensed fishermen complain about the undersized fish being caught. If the fish are too large, the rate of feeding the

fish is too high (i.e., fish size increases with the amount of feed), it costs the state more to feed the fish than was built into the hatchery budget.
Let's suppose that the state wants the average weight of the trout released into streams and lakes from a fish hatchery to be 308 grams (g) or 11 ounces (oz.). (Note: There are 28 g in one ounce.) Suppose you have been asked to analyze the hypothetical data in Fig. 3.11 which gives the weight of a random sample of trout released during the past week from a Colorado hatchery. The hypothetical data are given in Fig. 3.11.

WEIGHT OF TROUT WHEN RELEASED FROM FISH HATCHERY
Weight (g)
254.8
291.2
324.8
294.0
355.6
347.2
347.2
347.2
305.2
313.6
366.8
350.0
366.8
361.2

Fig. 3.11 Worksheet Data for Chap. 3: Practice Problem #2

Create an Excel spreadsheet with these data.

(a) Use Excel to the right of the table to find the sample size, mean, standard deviation, and standard error of the mean for these data. Label your answers, and use two decimal places for the mean, standard deviation, and standard error of the mean
(b) Enter the null hypothesis and the research hypothesis for this problem on your spreadsheet.
(c) Use Excel's TINV function to find the 95 % confidence interval about the mean for these data. Label your answers on your spreadsheet. Use two decimal places for the lower limit and the upper limit of the confidence interval.
(d) Enter the *result* of the test on your spreadsheet.

(e) Enter the *conclusion* of the test in plain English on your spreadsheet.
(f) Print your final spreadsheet so that it fits onto one page (if you need help remembering how to do this, see the objectives at the end of Chap. 2 in Sect. 2.4).
(g) Draw a picture of the confidence interval, including the reference value, onto your spreadsheet.
(h) Save the final spreadsheet as: TROUT10

3. According to Bremer and Doerge (2010), only one species of oak tree, *Quercus tomentella Engelm*, grows on the island of Guadalupe. Let's suppose that 5 years ago, the average size of acorn nut as measured by length from this species of oak was 27 millimeters (mm). Suppose that you have been asked to determine if this average size has changed this year. You decide to run a small sample of oak acorns to test your Excel skills, and the hypothetical data are given in Fig. 3.12:

LENGTH OF ACORNS FROM OAK TREES

Length (mm)
20
18
22
23
27
26
28
29
30
30
33
26
27
28
38
37
30
32
33
31
37
38

Fig. 3.12 Worksheet Data for Chap. 3: Practice Problem #3

Create an Excel spreadsheet with these data.

(a) Use Excel to the right of the table to find the sample size, mean, standard deviation, and standard error of the mean for these data. Label your answers, and use two decimal places for the mean, standard deviation, and standard error of the mean
(b) Enter the null hypothesis and the research hypothesis for this problem onto your spreadsheet.
(c) Use Excel's TINV function to find the 95 % confidence interval about the mean for these data. Label your answers on your spreadsheet. Use two decimal places for the lower limit and the upper limit of the confidence interval.
(d) Enter the *result* of the test on your spreadsheet.
(e) Enter the *conclusion* of the test in plain English on your spreadsheet.
(f) Print your final spreadsheet so that it fits onto one page (if you need help remembering how to do this, see the objectives at the end of Chap. 2 in Sect. 2.4).
(g) Draw a picture of the confidence interval, including the reference value, onto your spreadsheet.
(h) Save the final spreadsheet as: acorn10

References

Black, K. Business Statistics: for Contemporary Decision Making (6th ed.). Hoboken, NJ: John Wiley& Sons, Inc., 2010.

Bremer, M. and Doerge, R.W. Statistics at the Bench: A Step-by-Step Handbook for Biologists. Cold Spring Harbor, NY: Cold Spring Harbor Laboratory Press, 2010.

Keller, G. Statistics for Management and Economics (8th ed.). Mason, OH: South-Western Cengage learning, 2009.

Levine, D.M. Statistics for Managers using Microsoft Excel (6th ed.). Boston, MA: Prentice Hall/ Pearson, 2011.

McDaniel, C. and Gates, R. Marketing Research (8th ed.). Hoboken, NJ: John Wiley & Sons, Inc., 2010.

Salkind, N.J. Statistics for People Who (think they) Hate Statistics (2nd Excel 2007 ed.). Los Angeles, CA: Sage Publications, 2010.

Townend, J. Practical Statistics for Environmental and Biological Scientists. Hoboken, NJ: John Wiley & Sons, Inc., 2002.

van Emden, H.F. Statistics for Terrified Biologists. Malden, MA: Blackwell Publishing, 2008.

Weiers, R.M. Introduction to Business Statistics (7th ed.). Mason, OH: South-Western Cengage Learning, 2011.

Zikmund, W.G. and Babin, B.J. Exploring Marketing Research (10th ed.). Mason, OH: South-Western Cengage learning, 2010.

Chapter 4
One-Group t-Test for the Mean

In this chapter, you will learn how to use one of the most popular and most helpful statistical tests in science research: the one-group t-test for the mean.

The formula for the one-group t-test is as follows:

$$t = \frac{\overline{X} - \mu}{S_{\overline{X}}} \text{ where} \tag{4.1}$$

$$\text{s.e.} = S_{\overline{X}} = \frac{S}{\sqrt{n}} \tag{4.2}$$

This formula asks you to take the mean ($\overline{X}$) and subtract the population mean (μ) from it, and then divide the answer by the standard error of the mean (s.e.). The standard error of the mean equals the standard deviation divided by the square root of n (the sample size). If you want to learn more about this test, see Zikmund and Babin (2010).

Let's discuss the 7 STEPS of hypothesis testing using the one-group t-test so that you can understand how this test is used.

4.1 The Seven STEPS for Hypothesis-Testing Using the One-Group t-Test

Objective:	To learn the seven steps of hypothesis-testing using the one-group t-test

Before you can try out your Excel skills on the one-group t-test, you need to learn the basic steps of hypothesis-testing for this statistical test. There are seven steps in this process:

T.J. Quirk et al., *Excel 2016 for Biological and Life Sciences Statistics*, Excel for Statistics, DOI 10.1007/978-3-319-39489-3_4

4.1.1 STEP 1: State the Null Hypothesis and the Research Hypothesis

If you are using numerical scales in your survey, you need to remember that these hypotheses refer to the "middle" of the numerical scale. For example, if you are using 7-point scales with 1 = poor and 7 = excellent, these hypotheses would refer to the middle of these scales and would be:

Null hypothesis H_0: $\mu = 4$
Research hypothesis H_1: $\mu \neq 4$

As a second example, suppose that you worked for Honda Motor Company and that you wanted to place a magazine ad that claimed that the new Honda Fit got 35 miles per gallon (mpg). The hypotheses for testing this claim on actual data would be:

H_0: $\mu = 35\,\text{mpg}$
H_1: $\mu \neq 35\,\text{mpg}$

4.1.2 STEP 2: Select the Appropriate Statistical Test

In this chapter we will be studying the one-group t-test, and so we will select that test.

4.1.3 STEP 3: Decide on a Decision Rule for the One-Group t-Test

(a) If the absolute value of t is less than the critical value of t, accept the null hypothesis.
(b) If the absolute value of t is greater than the critical value of t, reject the null hypothesis and accept the research hypothesis.

You are probably saying to yourself: "That sounds fine, but how do I find the absolute value of t?"

4.1.3.1 Finding the Absolute Value of a Number

To do that, we need another objective:

Objective: To find the absolute value of a number

If you took a basic algebra course in high school, you may remember the concept of "absolute value." In mathematical terms, the absolute value of any number is *always* that number expressed as a positive number.

For example, the absolute value of 2.35 is +2.35.

And the absolute value of minus 2.35 (i.e. −2.35) is also +2.35.

This becomes important when you are using the t-table in Appendix E of this book. We will discuss this table later when we get to Step 5 of the one-group t-test where we explain how to find the critical value of t using Appendix E.

4.1.4 STEP 4: Calculate the Formula for the One-Group t-Test

Objective: To learn how to use the formula for the one-group t-test

The formula for the one-group t-test is as follows:

$$t = \frac{\overline{X} - \mu}{S_{\overline{X}}} \text{ where} \tag{4.1}$$

$$\text{s.e.} = S_{\overline{X}} = \frac{S}{\sqrt{n}} \tag{4.2}$$

This formula makes the following assumptions about the data (Foster, Stine, and Waterman 1998): (1) The data are independent of each other (i.e., each person receives only one score), (2) the *population* of the data is normally distributed, and (3) the data have a constant variance (note that the standard deviation is the square root of the variance).

To use this formula, you need to follow these steps:

1. Take the sample mean in your research study and subtract the population mean μ from it (remember that the population mean for a study involving numerical rating scales is the "middle" number in the scale).
2. Then take your answer from the above step, and divide your answer by the standard error of the mean for your research study (you will remember that you learned how to find the standard error of the mean in Chap. 1; to find the standard error of the mean, just take the standard deviation of your research study and divide it by the square root of n, where n is the number of people, plants, or animals used in your research study).
3. The number you get after you complete the above step is the value for t that results when you use the formula stated above.

4.1.5 STEP 5: Find the Critical Value of t in the t-Table in Appendix E

Objective: To find the critical value of t in the t-table in Appendix E

Before we get into an explanation of what is meant by "the critical value of t," let's give you practice in finding the critical value of t by using the t-table in Appendix E.

Keep your finger on Appendix E as we explain how you need to "read" that table.

Since the test in this chapter is called the "one-group t-test," you will use the first column on the left in Appendix E to find the critical value of t for your research study (note that this column is headed: "sample size n").

To find the critical value of t, you go down this first column until you find the sample size in your research study, and then you go to the right and read the critical value of t for that sample size in the critical t column in the table (note that *this is the column that you would use for both the one-group t-test and the 95 % confidence interval about the mean*).

For example, if you have 27 people in your research study, the critical value of t is 2.056.

If you have 38 people in your research study, the critical value of t is 2.026.

If you have more than 40 people in your research study, the critical value of t is always 1.96.

Note that the "critical t column" in Appendix E represents the value of t that you need to obtain to be 95 % confident of your results as "significant" results.

The critical value of t is the value that tells you whether or not you have found a "significant result" in your statistical test.

The t-table in Appendix E represents a series of "bell-shaped normal curves" (they are called bell-shaped because they look like the outline of the Liberty Bell that you can see in Philadelphia outside of Independence Hall).

The "middle" of these normal curves is treated as if it were zero point on the x-axis (the technical explanation of this fact is beyond the scope of this book, but any good statistics book (e.g., Zikmund and Babin 2010) will explain this concept to you if you are interested in learning more about it).

Thus, values of t that are to the right of this zero point are positive values that use a plus sign before them, and values of t that are to the left of this zero point are negative values that use a minus sign before them. Thus, some values of t are positive, and some are negative.

However, every statistics book that includes a t-table only reprints the *positive* side of the t-curves because the negative side is the mirror image of the positive side; this means that the negative side contains the exact same numbers as the positive side, but the negative numbers all have a minus sign in front of them.

Therefore, to use the t-table in Appendix E, you need to *take the absolute value of the t-value you found when you use the t-test formula* since the t-table in Appendix E only has the values of t that are the positive values for t.

Throughout this book, we are assuming that you want to be 95 % confident in the results of your statistical tests. Therefore, the value for t in the t-table in Appendix E tells you whether or not the t-value you obtained when you used the formula for the one-group t-test is within the 95 % interval of the t-curve range that that t-value would be expected to occur with 95 % confidence.

If the t-value you obtained when you used the formula for the one-group t-test is *inside* of the 95 % confidence range, we say that the result you found is *not significant* (note that this is equivalent to *accepting the null hypothesis!*).

If the t-value you found when you used the formula for the one-group t-test is *outside* of this 95 % confidence range, we say that you have found a *significant result* that would be expected to occur less than 5 % of the time (note that this is equivalent to *rejecting the null hypothesis and accepting the research hypothesis*).

4.1.6 STEP 6: State the Result of Your Statistical Test

There are two possible results when you use the one-group t-test, and only one of them can be accepted as "true."

Either: Since the absolute value of t that you found in the t-test formula is *less than the critical value of t* in Appendix E, you accept the null hypothesis.

Or: Since the absolute value of t that you found in the t-test formula is *greater than the critical value of t* in Appendix E, you reject the null hypothesis, and accept the research hypothesis.

4.1.7 STEP 7: State the Conclusion of Your Statistical Test in Plain English!

In practice, this is more difficult than it sounds because you are trying to summarize the result of your statistical test in simple English that is both concise and accurate so that someone who has never had a statistics course (such as your boss, perhaps) can understand the result of your test. This is a difficult task, and we will give you lots of practice doing this last and most important step throughout this book.

If you have read this far, you are ready to sit down at your computer and perform the one-group t-test using Excel on some hypothetical data.

Let's give this a try.

4.2 One-Group t-Test for the Mean

The American Institute of Biological Sciences (AIBS) is a non-profit scientific association dedicated to advancing biological research and education. It has more than 150 member organizations. AIBS publishes a peer-reviewed monthly journal called *BioScience*. Its Website links to Career Possibilities for biologists in academics, government, and private industry. Let's suppose that AIBS has sent an Internet survey to its members asking them to evaluate its Website in terms of its career possibilities section.

The survey contains a number of items, but suppose a hypothetical Item #7 is the one in Fig. 4.1:

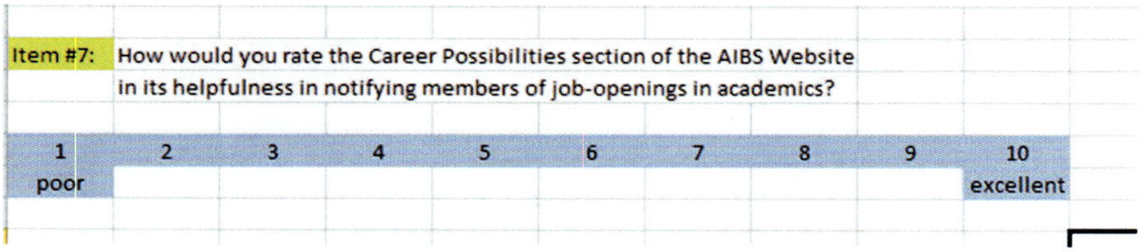

Fig. 4.1 Sample Survey Item for Item #7 of the AIBS Survey (Practical Example)

Suppose further, that you have decided to analyze the data from members using the one-group t-test.

Important note: You would need to use this test for each of the survey items separately.

Suppose that the hypothetical data for Item #7 of the AIBS Website were based on a sample size of 124 members who had a mean score on this item of 6.58 and a standard deviation on this item of 2.44.

Objective: To analyze the data for each question separately using the one-group t-test for each survey item.

Create an Excel spreadsheet with the following information:

B11: Null hypothesis:
B14: Research hypothesis:

Note: Remember that when you are using a rating scale item, both the null hypothesis and the research hypothesis refer to the "middle of the scale." In the 10-point scale in this example, the middle of the scale is 5.5 since five numbers are below 5.5 (i.e., 1-5) and five numbers are above 5.5 (i.e. 6–10). Therefore, the hypotheses for this rating scale item are:

H_0: $\mu = 5.5$
H_1: $\mu \neq 5.5$

B17: n
B20: mean
B23: STDEV

B26: s.e.
B29: critical t
B32: t-test
B36: Result:
B41: Conclusion:

Now, use Excel:

D17: enter the sample size
D20: enter the mean
D23: enter the STDEV (see Fig. 4.2)

Fig. 4.2 Basic Data Table for Item #7 of the AIBS Survey

Null hypothesis:		
Research hypothesis:		
n		124
mean		6.58
STDEV		2.44
s.e.		
critical t		
t-test		
Result:		
Conclusion:		

D26: compute the standard error using the formula in Chap. 1
D29: find the critical t value of t in the t-table in Appendix E

Now, enter the following formula in cell D32 to find the t-test result:

= (D20 – 5.5)/D26 (no spaces between symbols)

This formula takes the sample mean (D20) and subtracts the population hypothesized mean of 5.5 from the sample mean, and THEN divides the answer by the standard error of the mean (D26). Note that you need to enter D20–5.5 with an open-parenthesis *before* D20 and a closed-parenthesis *after* 5.5 so that the *answer of 1.08 is THEN divided by the standard error of 0.22* to get the t-test result of 4.93.

Now, use two decimal places for both the s.e. and the t-test result (see Fig. 4.3).

Null hypothesis:		
Research hypothesis:		
n		124
mean		6.58
STDEV		2.44
s.e.		0.22
critical t		1.96
t-test		4.93
Result:		
Conclusion:		

Fig. 4.3 t-test Formula Result for Item #7 of the AIBS Survey

Now, write the following sentence in D36-D39 to summarize the result of the t-test:

D36: Since the absolute value of t of 4.93 is
D37: greater than the critical t of 1.96, we
D38: reject the null hypothesis and accept
D39: the research hypothesis.

Lastly, write the following sentence in D41-D44 to summarize the conclusion of the result for Item #7 of the AIBS Survey:

D41: Members rated the helpfulness of the Career
D42: Possibilities section of the AIBS Website in its
D43: helpfulness in notifying members of job-openings
D44: in academics as significantly positive.

Save your file as: career4

Important note: We have used the term "significantly positive" because the mean rating of 6.58 is on the positive side of the rating scale. We purposely have not used the term "significantly excellent" because people who speak English do not use that term because something is either excellent or it is not excellent. Therefore, "significantly positive" is a more correct use of the English language in this type of rating scale item.

Important note: You are probably wondering why we entered both the result and the conclusion in separate cells instead of in just one cell. This is because if you enter them in one cell, you will be very disappointed when you print out your final spreadsheet, because one of two things will happen that you will not like: (1) if you print the spreadsheet to fit onto only one page, the result and the conclusion will force the entire spreadsheet to be printed in such small font size that you will be unable to read it, or (2) if you do not print the final spreadsheet to fit onto one page, both the result and the conclusion will "dribble over" onto a second page instead of fitting the entire spread-sheet onto one page. In either case, your spread-sheet will not have a "professional look."

Print the final spreadsheet so that it fits onto one page as given in Fig. 4.4. Enter the null hypothesis and the research hypothesis by hand on your spreadsheet

Null hypothesis:		μ	=	5.5		
Research hypothesis:		μ	≠	5.5		
n		124				
mean		6.58				
STDEV		2.44				
s.e.		0.22				
critical t		1.96				
t-test		4.93				
Result:		Since the absolute value of t of 4.93 is greater than the critical t of 1.96, we reject the null hypothesis and accept the research hypothesis.				
Conclusion:		Members rated the helpfulness of the Career Possibilities section of the AIBS Website in its helpfulness in notifying members of job-openings in academics as significantly positive.				

Fig. 4.4 Final Spreadsheet for Item #7 of the AIBS Survey

Important Note: It is important for you to understand that "technically" the above conclusion in statistical terms should state:

"Members rated the helpfulness of the Career Possibilities section of the AIBS Website in its helpfulness in notifying members of job-openings in academics as positive, and this result was probably not obtained by chance."

However, throughout this book, we are using the term "significantly" in writing the conclusion of statistical tests to alert the reader that the result of the statistical test was probably not a chance finding, but instead of writing all of those words each time, we use the word "significantly" as a shorthand to the longer explanation. This makes it much easier for the reader to understand the conclusion when it is written "in plain English," instead of technical, statistical language.

4.3 Can You Use Either the 95 % Confidence Interval About the Mean OR the One-Group t-Test When Testing Hypotheses?

You are probably asking yourself:

"It sounds like you could use *either* the 95 % confidence interval about the mean *or* the one-group t-test to analyze the results of the types of problems described so far in this book? Is this a correct statement?"

The answer is a resounding: *"Yes!"*

Both the confidence interval about the mean and the one-group t-test are used often in science research on the types of problems described so far in this book. *Both of these tests produce the same result and the same conclusion from the data set!*

Both of these tests are explained in this book because some researchers prefer the confidence interval about the mean test, others prefer the one-group t-test, and still others prefer to use both tests on the same data to make their results and conclusions clearer to the reader of their research reports. Since we do not know which of these tests your researcher prefers, we have explained both of them so that you are competent in the use of both tests in the analysis of statistical data.

Now, let's try your Excel skills on the one-group t-test on these three problems at the end of this chapter.

4.4 End-of-Chapter Practice Problems

1. Suppose that you have been asked to decide if the average petal length of a particular plant species grown in the state of Missouri has changed from what it was 5 years ago when the average length of this plant in selected sites in Missouri was 3.10 centimeters (cm). You have been asked to test your Excel skills on the hypothetical data given in Fig. 4.5.

Fig. 4.5 Worksheet Data for Chap. 4: Practice Problem #1

PETAL LENGTH OF A PLANT SPECIES
Length (in cm)
4.72
4.54
4.61
4.23
4.12
3.97
3.85
3.74
3.65
3.55
3.12
2.97
2.86
2.71
2.64
2.56

(a) Write the null hypothesis and the research hypothesis on your spreadsheet
(b) Use Excel to find the sample size, mean, standard deviation, and standard error of the mean to the right of the data set. Use number format (two decimal places) for the mean, standard deviation, and standard error of the mean.
(c) Enter the critical t from the t-table in Appendix E onto your spreadsheet, and label it.
(d) Use Excel to compute the t-value for these data (use two decimal places) and label it on your spreadsheet
(e) Type the result on your spreadsheet, and then type the conclusion in plain English on your spreadsheet
(f) Save the file as: PETAL3

2. Suppose that you wanted to study the mass (in grams) of rainbow trout (*Oncorhynchus mykiss*) in a river in southern Colorado in the U.S. Five years ago, the average mass was 112 grams (g). You would like to know if the average mass has changed since then. You have collected data on a sample of rainbow trout from the same river, and you want to test your Excel skills on a small

sample before you try to do the larger data analysis. The hypothetical data for a random sample of rainbow trout is presented in Fig. 4.6:

Fig. 4.6 Worksheet Data for Chap. 4: Practice Problem #2

COLORADO RAINBOW TROUT MASS (in grams)
MASS (grams)
114
110
117
112
115
116
112
125
118
113
120
112
114
117
119

(a) *On your Excel spreadsheet*, write the null hypothesis and the research hypothesis for these data.
(b) Use Excel to find the *sample size, mean, standard deviation, and standard error of the mean* for these data (two decimal places for the mean, standard deviation, and standard error of the mean).
(c) Use Excel to perform a *one-group t-test* on these data (two decimal places).
(d) On your printout, type the *critical value of t* given in your t-table in Appendix E.
(e) On your spreadsheet, type the *result* of the t-test.
(f) On your spreadsheet, type the *conclusion* of your study in plain English.
(g) save the file as: TROUT33

3. Suppose that you wanted to study the mating call of a tree frog, *Hyla versicolor*, by measuring the duration in milliseconds (msec) of the notes. In a previous study 8 years ago, the average length of the mating call in this type of frog was 160 msec. You have collected data and want to test your Excel skills on a small sample before you try to analyze the data from a much larger sample. The hypothetical data appear in Fig. 4.7.

Fig. 4.7 Worksheet Data for Chap. 4: Practice Problem #3

TREE FROG MATING CALL DURATION
DURATION (in msec)
100
120
135
147
180
180
220
295
287
289
300
295
276
268
254
264
248
296

(a) Write the null hypothesis and the research hypothesis on your spreadsheet
(b) Use Excel to find the sample size, mean, standard deviation, and standard error of the mean to the right of the data set. Use number format (two decimal places) for the mean, standard deviation, and standard error of the mean.
(c) Enter the critical t from the t-table in Appendix E onto your spreadsheet, and label it.
(d) Use Excel to compute the t-value for these data (use two decimal places) and label it on your spreadsheet
(e) Type the result on your spreadsheet, and then type the conclusion in plain English on your spreadsheet
(f) Save the file as: FROG3

References

Zikmund, W.G. and Babin, B.J. Exploring Marketing Research (10th ed.) Mason, OH: South-Western Cengage Learning, 2010.

Foster, D.P., Stine, R.A., Waterman, R.P. Basic Business Statistics: A Casebook. New York, NY: Springer-Verlag, 1998.

Chapter 5
Two-Group t-Test of the Difference of the Means for Independent Groups

Up until now in this book, you have been dealing with the situation in which you have had only one group of people, plants, or animals in your research study and only one measurement "number" on each of these people, plants, or animals. We will now change gears and deal with the situation in which you are measuring two groups instead of only one group.

Whenever you have two completely different groups of people (i.e., no one person is in both groups, but every person is measured on only one variable to produce one "number" for each person), we say that the two groups are "independent of one another." This chapter deals with just that situation and that is why it is called the two-group t-test for independent groups.

The two assumptions underlying the two-group t-test are the following (Zikmund and Babin 2010): (1) both groups are sampled from a normal population, and (2) the variances of the two populations are approximately equal. Note that the standard deviation is merely the square root of the variance. (There are different formulas to use when each person is measured twice to create two groups of data, and this situation is called "dependent," but those formulas are beyond the scope of this book.) This book only deals with two groups that are independent of one another so that no person is in both groups of data.

When you are testing for the difference between the means for two groups, it is important to remember that there are two different formulas that you need to use depending on the sample sizes of the two groups:

(1) Use Formula #1 in this chapter when both of the groups have a sample size greater than 30, and
(2) Use Formula #2 in this chapter when either one group, or both groups, have a sample size less than 30.

We will illustrate both of these situations in this chapter.

But, first, we need to understand the steps involved in hypothesis-testing when two groups are involved before we dive into the formulas for this test.

T.J. Quirk et al., *Excel 2016 for Biological and Life Sciences Statistics*, Excel for Statistics, DOI 10.1007/978-3-319-39489-3_5

5.1 The 9 STEPS for Hypothesis-Testing Using the Two-Group t-Test

Objective: To learn the 9 steps of hypothesis-testing using two groups of people, plants, or animals and the two-group t-test

You will see that these steps parallel the steps used in the previous chapter that dealt with the one-group t-test, but there are some important differences between the steps that you need to understand clearly before we dive into the formulas for the two-group t-test.

5.1.1 STEP 1: Name One Group, Group 1, and the Other Group, Group 2

The formulas used in this chapter will use the numbers 1 and 2 to distinguish between the two groups. If you define which group is Group 1 and which group is Group 2, you can use these numbers in your computations without having to write out the names of the groups.

For example, if you are testing the perceived effectiveness of two different types of pain medications on college freshmen males, you could call the groups: "Ibuprofen" and "Acetaminophen," but this would require your writing out the words "Ibuprofen" and Acetaminophen" whenever you wanted to refer to one of these groups. If you call the Ibuprofen group, Group 1, and the Acetaminophen group, Group 2, this makes it much easier to refer to the groups because it saves you writing time.

As a second example, you could be comparing flower preference for one type of hummingbird. Two types of flowers, fuschias and mandevillas, have a vibrant red color that is a typical attractant found in hummingbird gardens. If you had to write out the names of the two flowers whenever you wanted to refer to them, it would take you more time than it would if, instead, you named one flower, Group 1, and the other flower, Group 2.

Note, also, that it is completely arbitrary which group you call Group 1, and which Group you call Group 2. You will achieve the same result and the same conclusion from the formulas however you decide to define these two groups.

5.1.2 STEP 2: Create a Table That Summarizes the Sample Size, Mean Score, and Standard Deviation of Each Group

This step makes it easier for you to make sure that you are using the correct numbers in the formulas for the two-group t-test. If you get the numbers "mixed-up," your entire formula work will be incorrect and you will botch the problem terribly.

For example, suppose that you tested college freshmen males on the perceived effectiveness of "Ibuprofen" versus "Acetaminophen," in which the males were randomly assigned to use just one of these types of pain medication, and then to rate its perceived effectiveness on a 100-point scale from 0 = poor to 100 = excellent. After the research study was completed, suppose that the Ibuprofen group had 52 males in it, their mean effectiveness rating was 55 with a standard deviation of 7, while the Acetaminophen group had 57 males in it and their average effectiveness rating was 64 with a standard deviation of 13.

The formulas for analyzing these data to determine if there was a significant difference in the effectiveness rating for freshmen males for these two types of pain medication require you to use six numbers correctly in the formulas: the sample size, the mean, and the standard deviation of each of the two groups. All six of these numbers must be used correctly in the formulas if you are to analyze the data correctly.

If you create a table to summarize these data, a good example of the table, using both Step 1 and Step 2, would be the data presented in Fig. 5.1:

	A	B	C	D	E	F
1						
2						
3		Group	n	Mean	STDEV	
4		1 (name it)				
5		2 (name it)				
6						
7						

Fig. 5.1 Basic Table Format for the Two-group t-test

For example, if you decide to call Group 1 the Ibuprofen group and Group 2 the Acetaminophen group, the following table would place the six numbers from your research study into the proper calls of the table as in Fig. 5.2:

	A	B	C	D	E	F
1						
2						
3		Group	n	Mean	STDEV	
4		1 Ibuprofen	52	55	7	
5		2 Acetaminophen	57	64	13	
6						
7						

Fig. 5.2 Results of Entering the Data Needed for the Two-group t-test

You can now use the formulas for the two-group t-test with more confidence that the six numbers will be placed in the proper place in the formulas.

Note that you could just as easily call Group 1 the Acetaminophen group and Group 2 the Ibuprofen group; it makes no difference how you decide to name the two groups; this decision is up to you.

5.1.3 STEP 3: State the Null Hypothesis and the Research Hypothesis for the Two-Group t-Test

If you have completed Step 1 above, this step is very easy because the null hypothesis and the research hypothesis will always be stated in the same way for the two-group t-test. The null hypothesis states that the population means of the two groups are equal, while the research hypothesis states that the population means of the two groups are not equal. In notation format, this becomes:

H_0: $\mu_1 = \mu_2$
H_1: $\mu_1 \neq \mu_2$

You can now see that this notation is much simpler than having to write out the names of the two groups in all of your formulas.

5.1.4 STEP 4: Select the Appropriate Statistical Test

Since this chapter deals with the situation in which you have two groups but only one measurement on each person, plant, or animal in each group, we will use the two-group t-test throughout this chapter.

5.1.5 STEP 5: Decide on a Decision Rule for the Two-Group t-Test

The decision rule is exactly what it was in the previous chapter (see Sect. 4.1.3) when we dealt with the one-group t-test.

(a) If the absolute value of t is less than the critical value of t, accept the null hypothesis.
(b) If the absolute value of t is greater than the critical value of t, reject the null hypothesis and accept the research hypothesis.

Since you learned how to find the absolute value of t in the previous chapter (see Sect. 4.1.3.1), you can use that knowledge in this chapter.

5.1.6 STEP 6: Calculate the Formula for the Two-Group t-Test

Since we are using two different formulas in this chapter for the two-group t-test depending on the sample size in the two groups, we will explain how to use those formulas later in this chapter.

5.1.7 STEP 7: Find the Critical Value of t in the t-Table in Appendix E

In the previous chapter where we were dealing with the one-group t-test, you found the critical value of t in the t-table in Appendix E by finding the sample size for the one group in the first column of the table, and then reading the critical value of t across from it on the right in the "critical t column" in the table (see Sect. 4.1.5). This process was fairly simple once you have had some practice in doing this step.

However, for the two-group t-test, the procedure for finding the critical value of t is more complicated because you have two different groups in your study, and they often have different sample sizes in each group.

To use Appendix E correctly in this chapter, you need to learn how to find the "degrees of freedom" for your study. We will discuss that process now.

5.1.7.1 Finding the Degrees of Freedom (df) for the Two-Group t-Test

Objective: To find the degrees of freedom for the two-group t-test and to use it to find the critical value of t in the t-table in Appendix E

The mathematical explanation of the concept of the "degrees of freedom" is beyond the scope of this book, but you can find out more about this concept by reading any good statistics book (e.g. Keller 2009). For our purposes, you can easily understand how to find the degrees of freedom and to use it to find the critical value of t in Appendix E. The formula for the degrees of freedom (df) is:

$$\text{degrees of freedom} = \text{df} = n_1 + n_2 - 2 \tag{5.1}$$

In other words, you add the sample size for Group 1 to the sample size for Group 2 and then subtract 2 from this total to get the number of degrees of freedom to use in Appendix E.

Take a look at Appendix E.

Instead of using the first column as we did in the one-group t-test that is based on the sample size, n, of one group, we need to use the second-column of this table (df) to find the critical value of t for the two-group t-test.

For example, if you had 13 people in Group 1 and 17 people in Group 2, the degrees of freedom would be: $13 + 17 - 2 = 28$, and the critical value of t would be 2.048 *since you look down the second column which contains the degrees of freedom* until you come to the number 28, and then read 2.048 in the "critical t column" in the table to find the critical value of t when $\text{df} = 28$.

As a second example, if you had 52 people in Group 1 and 57 people in Group 2, the degrees of freedom would be: $52 + 57 - 2 = 107$. When you go down the second column in Appendix E for the degrees of freedom, you find that *once you go beyond the degrees of freedom equal to 39, the critical value of t is always 1.96*, and that is the value you would use for the critical t with this example.

5.1.8 STEP 8: State the Result of Your Statistical Test

The result follows the exact same result format that you found for the one-group t-test in the previous chapter (see Sect. 4.1.6):

Either: Since the absolute value of t that you found in the t-test formula is *less than the critical value of t* in Appendix E, you accept the null hypothesis.

Or: Since the absolute value of t that you found in the t-test formula is *greater than the critical value of t* in Appendix E, you reject the null hypothesis and accept the research hypothesis.

5.1.9 STEP 9: State the Conclusion of Your Statistical Test in Plain English!

Writing the conclusion for the two-group t-test is more difficult than writing the conclusion for the one-group t-test because you have to decide what the difference was between the two groups.

When you accept the null hypothesis, the conclusion is simple to write: "There is no difference between the two groups in the variable that was measured."

But when you reject the null hypothesis and accept the research hypothesis, you need to be careful about writing the conclusion so that it is both accurate and concise.

Let's give you some practice in writing the conclusion of a two-group t-test.

5.1.9.1 Writing the Conclusion of the Two-Group t-Test When You Accept the Null Hypothesis

Objective: To write the conclusion of the two-group t-test when you have accepted the null hypothesis.

Suppose that a large state university wanted to study the satisfaction of graduates who had been Biology majors with their program for alumni who had graduated between 5 and 10 years ago A survey has been developed, and tried out with a pilot study with just a few graduates to see how it was working. Item #10 of this survey is given in Fig. 5.3 with some hypothetical data.

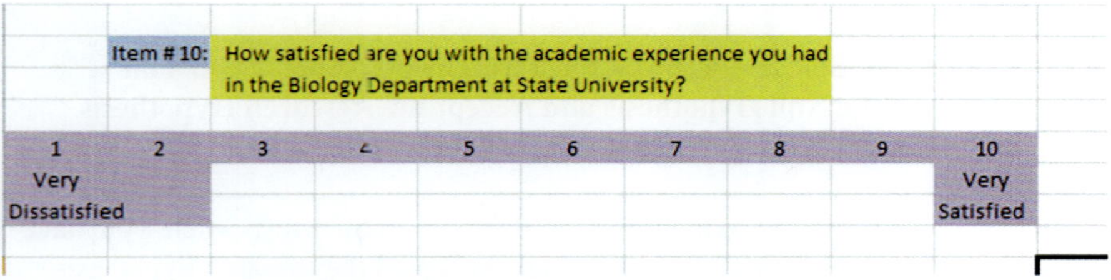

Fig. 5.3 State University Satisfaction Survey Item #10

Suppose further, that you have decided to analyze the data from alumni who were Biology majors by comparing men and women using the two-group t-test.

Important note: You would need to use this test for each of the survey items separately.

Suppose that the hypothetical data for Item #10 was based on a sample size of 124 men who had a mean score on this item of 6.58 and a standard deviation on this item of 2.44. Suppose that you also had data from 86 women who had a mean score of 6.45 with a standard deviation of 1.86.

We will explain later in this chapter how to produce the results of the two-group t-test using its formulas, but, for now, let's "cut to the chase" and tell you that those formulas would produce the following in Fig. 5.4:

Group	n	Mean	STDEV
1 Males	124	6.58	2.44
2 Females	86	6.45	1.86

Fig. 5.4 Worksheet Data for Males vs. Females for Item #10 for Accepting the Null Hypothesis

degrees of freedom:	208
critical t:	1.96 (in Appendix E)
t-test formula:	0.44 (when you use your calculator!)
Result:	Since the absolute value of 0.44 is less than the critical t of 1.96, we accept the null hypothesis.
Conclusion:	There was no difference between male and female alumni who were Biology majors in their satisfaction with the academic experience they had at State University.

Now, let's see what happens when you reject the null hypothesis (H_0) and accept the research hypothesis (H_1).

5.1.9.2 Writing the Conclusion of the Two-Group t-Test When You Reject the Null Hypothesis and Accept the Research Hypothesis

> Objective: To write the conclusion of the two-group t-test when you have rejected the null hypothesis and accepted the research hypothesis

Let's continue with this same example, but with the result that we reject the null hypothesis and accept the research hypothesis.

Let's assume that this time you have alumni data on 85 males and their mean score on this question was 7.26 with a standard deviation of 2.35. Let's further suppose that you also have data on 48 females and their mean score on this question was 4.37 with a standard deviation of 3.26.

Without going into the details of the formulas for the two-group t-test, these data would produce the following result and conclusion based on Fig. 5.5:

	A	B	C	D	E	F
1						
2						
3		Group	n	Mean	STDEV	
4		1 Males	85	7.26	2.35	
5		2 Females	48	4.37	3.26	
6						

Fig. 5.5 Worksheet Data for Item #10 for Obtaining a Significant Difference between Males and Females

Null Hypothesis:	$\mu_1 = \mu_2$
Research Hypothesis:	$\mu_1 \neq \mu_2$
degrees of freedom:	131
critical t:	1.96 (in Appendix E)
t-test formula:	5.40 (when you use your calculator!)
Result:	Since the absolute value of 5.40 is greater than the critical t of 1.96, we reject the null hypothesis and accept the research hypothesis.

Now, you need to compare the ratings of the men and women to find out which group had the more positive rating of their academic experience using the following rule:

Rule: To summarize the conclusion of the two-group t-test, just compare the means of the two groups, and be sure to use the word "significantly" in your conclusion if you rejected the null hypothesis and accepted the research hypothesis.

A good way to prepare to write the conclusion of the two-group t-test when you are using a rating scale is to place the mean scores of the two groups on a drawing of the scale so that you can visualize the difference of the mean scores. For example, for our Biology majors alumni example above, you would draw this "picture" of the scale in Fig. 5.6:

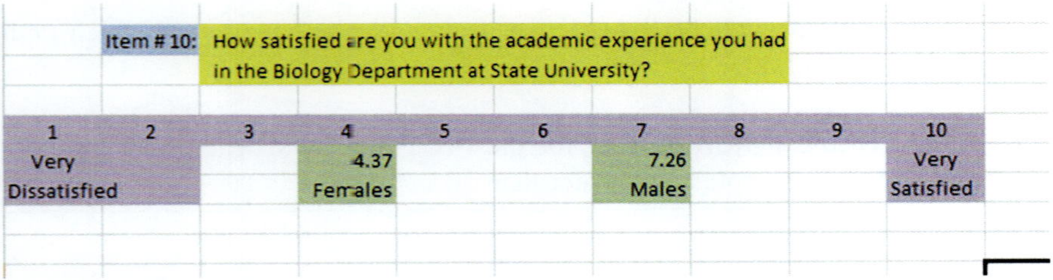

Fig. 5.6 Example of Drawing a "Picture" of the Means of the Two Groups on the Rating Scale

This drawing tells you visually that male alumni had a higher positive rating than female alumni on this item (7.26 vs. 4.37). *And, since you rejected the null hypothesis and accepted the research hypothesis, you know that you have found a significant difference between the two mean scores.*

So, our conclusion needs to contain the following key words:

- Male alumni
- Female alumni
- State University
- Biology majors
- significantly
- more satisfied ***or*** less satisfied
- *either* (7.26 vs. 4.37) *or* (4.37 vs. 7.26)

We can use these key words to write the either of two conclusions which are *logically identical*:

Either: Alumni who were male Biology majors were significantly more satisfied with their academic experience in the Biology department at State University than female Biology majors (7.26 vs. 4.37).

Or: Alumni who were female Biology majors were significantly less satisfied with their academic experience in the Biology department at State University than male Biology majors (4.37 vs. 7.26).

Both of these conclusions are accurate, so you can decide which one you want to write. It is your choice.

Also, note that the mean scores in parentheses at the end of these conclusions must match the sequence of the two groups in your conclusion. For example, if you say that: "Male alumni were significantly more satisfied than female alumni," the end of this conclusion should be: (7.26 vs. 4.37) since you mentioned males first, and females second.

Alternately, if you wrote that: "Female alumni were significantly less satisfied than male alumni," the end of this conclusion should be: (4.37 vs. 7.26) since you mentioned females first, and males second.

Putting the two mean scores at the end of your conclusion saves the reader from having to turn back to the table in your research report to find these mean scores to see how far apart the mean scores were.

Now, let's discuss FORMULA #1 that deals with the situation in which both groups have a sample size greater than 30.

Objective: To use FORMULA #1 for the two-group t-test when both groups have a sample size greater than 30

5.2 Formula #1: Both Groups Have a Sample Size Greater Than 30

The first formula we will discuss will be used when you have two groups with a sample size greater than 30 in each group and one measurement on each member in each group. This formula for the two-group t-test is:

$$t = \frac{\overline{X}_1 - \overline{X}_2}{S_{\overline{X}_1 - \overline{X}_2}} \tag{5.2}$$

$$\text{where } S_{\overline{X}_1 - \overline{X}_2} = \sqrt{\frac{S_1^2}{n_1} + \frac{S_2^2}{n_2}} \tag{5.3}$$

$$\text{and where degrees of freedom} = df = n_1 + n_2 - 2 \tag{5.1}$$

This formula looks daunting when you first see it, but let's explain some of the parts of this formula:

We have explained the concept of "degrees of freedom" earlier in this chapter, and so you should be able to find the degrees of freedom needed for this formula in order to find the critical value of t in Appendix E.

In the previous chapter, *the formula for the one-group t-test was the following*:

$$t = \frac{\overline{X} - \mu}{S_{\overline{X}}} \tag{4.1}$$

$$\text{where s.e.} = S_{\overline{X}} = \frac{S}{\sqrt{n}} \tag{4.2}$$

For the one-group t-test, you found the mean score and subtracted the population mean from it, and then divided the result by the standard error of the mean (s.e.) to get the result of the t-test. You then compared the t-test result to the critical value of t to see if you either accepted the null hypothesis, or rejected the null hypothesis and accepted the research hypothesis.

The two-group t-test requires a different formula because you have two groups, each with a mean score on some variable. You are trying to determine whether to accept the null hypothesis that the *population means of the two groups are equal* (in other words, there is no difference statistically between these two means), or whether the difference between the means of the two groups is "sufficiently large" that you would accept *that there is a significant difference* in the mean scores of the two groups.

The numerator of the two-group t-test asks you to find the difference of the means of the two groups:

$$\overline{X}_1 - \overline{X}_2 \tag{5.4}$$

The next step in the formula for the two-group t-test is to divide the answer you get when you subtract the two means by the standard error of the difference of the two means, and *this is a different standard error of the mean that you found for the one-group t-test because there are two means in the two-group t-test.*

The standard error of the mean when you have two groups is called the "standard error of the difference of the means". This formula looks less scary when you break it down into four steps:

1. Square the standard deviation of Group 1, and divide this result by the sample size for Group 1 (n_1).
2. Square the standard deviation of Group 2, and divide this result by the sample size for Group 2 (n_2).
3. Add the results of the above two steps to get a total score.
4. *Take the square root of this total score* to find the standard error of the difference of the means between the two groups, $S_{\overline{X}_1 - \overline{X}_2} = \sqrt{\frac{S_1^2}{n_1} + \frac{S_2^2}{n_2}}$

This last step is the one that gives students the most difficulty when they are finding this standard error using their calculator, because they are in such a hurry to get to the answer that they forget to carry the square root sign down to the last step, and thus get a larger number than they should for the standard error.

5.2.1 An Example of Formula #1 for the Two-Group t-Test

Now, let's use Formula #1 in a situation in which both groups have a sample size greater than 30.

Suppose that a researcher wanted to identify the preference of laboratory mice, males and females, for a certain type of bait. The researcher wanted to develop a type of bait that targeted female mice more than male mice in order to increase the effectiveness of a pest control agent. Both groups of mice, males and females, were given samples of the same bait. The researcher recorded the number of times each mouse in each group ate some of the bait over a period of time.

Suppose the researcher collected the data and determined (using her new Excel skills) that the 52 male mice in this study had a mean result of 55 with a standard deviation of 7, while the 57 female mice in this study had a mean result of 64 with a standard deviation of 13.

Note that the two-group t-test does not require that both groups have the same sample size. This is another way of saying that the two-group t-test is "robust" (a fancy term that statisticians like to use).

The data then produced the following table in Fig. 5.7:

	A	B	C	D	E
1					
2					
3		Group	n	Mean	STDEV
4		1 Males	52	55	7
5		2 Females	57	64	13

Fig. 5.7 Worksheet Data for Mouse Study

Create an Excel spreadsheet, and enter the following information:

B3: Group
B4: 1 Males
B5: 2 Females
C3: n
D3: Mean
E3: STDEV
C4: 52
D4: 55
E4: 7
C5: 57
D5: 64
E5: 13

Now, widen column B so that it is twice as wide as column A, and center the six numbers and their labels in your table (see Fig. 5.8)

	A	B	C	D	E
1					
2					
3		Group	n	Mean	STDEV
4		1 Males	52	55	7
5		2 Females	57	64	13
6					

Fig. 5.8 Results of Widening Column B and Centering the Numbers in the Cells

B8: Null hypothesis:
B10: Research hypothesis:

Since both groups have a sample size greater than 30, you need to use Formula #1 for the t-test for the difference of the means of the two groups.

Let's "break this formula down into pieces" to reduce the chance of making a mistake.

B13: STDEV1 squared/n1 (note that you square the standard deviation of Group 1, and then divide the result by the sample size of Group 1)
B16: STDEV2 squared/n2
B19: D13 + D16
B22: s.e.
B25: critical t
B28: t-test
B31: Result:
B36: Conclusion: (see Fig. 5.9)

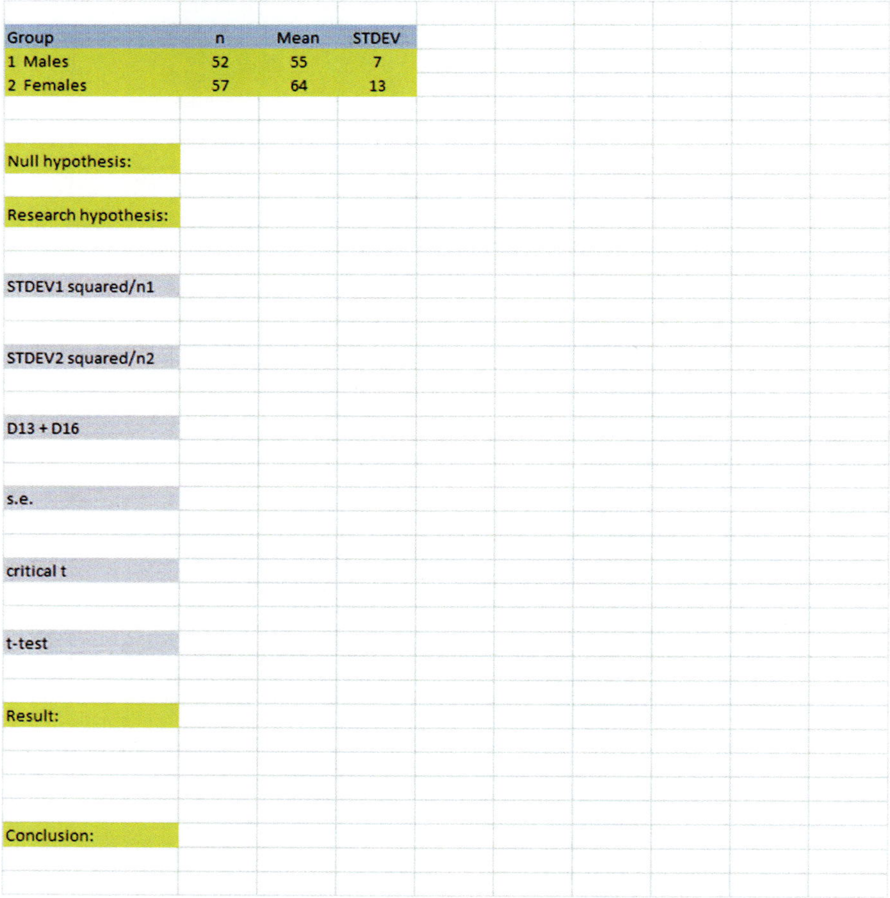

Group	n	Mean	STDEV
1 Males	52	55	7
2 Females	57	64	13

Null hypothesis:

Research hypothesis:

STDEV1 squared/n1

STDEV2 squared/n2

D13 + D16

s.e.

critical t

t-test

Result:

Conclusion:

Fig. 5.9 Formula Labels for the Two-group t-test

You now need to compute the values of the above formulas in the following cells:

D13: the result of the formula needed to compute cell B13 (use two decimals)
D16: the result of the formula needed to compute cell B16 (use two decimals)

D19: the result of the formula needed to compute cell B19 (use two decimals)
D22: =SQRT(D19) (use two decimals)

This formula should give you a standard error (s.e.) of 1.98.

D25: 1.96

(Since df = n1 + n2 – 2, this gives df = 109 – 2 = 107, and the critical t is, therefore, 1.96 in Appendix E.)

D28: =(D4 – D5)/D22 (use two decimals) (no spaces between symbols)

This formula should give you a value for the t-test of: −4.55.

Next, check to see if you have rounded off all figures in D13:D28 to two decimal places (see Fig. 5.10).

Group	n	Mean	STDEV
1 Males	52	55	7
2 Females	57	64	13
Null hypothesis:			
Research hypothesis:			
STDEV1 squared/n1		0.94	
STDEV2 squared/n2		2.96	
D13 + D16		3.91	
s.e.		1.98	
critical t		1.96	
t-test		−4.55	
Result:			
Conclusion:			

Fig. 5.10 Results of the t-test Formula for Mouse Study

Now, write the following sentence in D31 to D34 to summarize the result of the study:

D31: Since the absolute value of −4.55
D32: is greater than the critical t of
D33: 1.96, we reject the null hypothesis
D34: and accept the research hypothesis.

Finally, write the following sentence in D36 to D38 to summarize the conclusion of the study in plain English:

D36: Overall, the female mice significantly preferred
D37: the bait compared to the male mice (64 vs. 55)

Save your file as: MOUSE

Important note: You are probably wondering why we entered both the result and the conclusion in separate cells instead of in just one cell. This is because if you enter them in one cell, you will be very disappointed when you print out your final spreadsheet, because one of two things will happen that you will not like: (1) if you print the spreadsheet to fit onto only one page, the result and the conclusion will force the entire spreadsheet to be printed in such small font size that you will be unable to read it, or (2) if you do not print the final spreadsheet to fit onto one page, both the result and the conclusion will "dribble over" onto a second page instead of fitting the entire spreadsheet onto one page. In either case, your spreadsheet will not have a "professional look."

Print this file so that it fits onto one page, and write by hand the null hypothesis and the research hypothesis on your printout.

The final spreadsheet appears in Fig. 5.11.

Group	n	Mean	STDEV	
1 Males	52	55	7	
2 Females	57	64	13	
Null hypothesis:		μ_1	=	μ_2
Research hypothesis:		μ_1	≠	μ_2
STDEV1 squared/n1		0.94		
STDEV2 squared/n2		2.96		
D13 + D16		3.91		
s.e.		1.98		
critical t		1.96		
t-test		−4.55		
Result:		Since the absolute value of −4.55 is greater than the critical t of 1.96, we reject the null hypothesis and accept the research hypothesis.		
Conclusion:		Overall, the female mice significantly preferred the bait compared to the male mice (64 vs. 55).		

Fig. 5.11 Final Worksheet for Mouse Study

Now, let's use the second formula for the two-group t-test which we use whenever either one group, or both groups, have a sample size less than 30.

> Objective: To use Formula #2 for the two-group t-test when one or both groups have a sample size less than 30

Now, let's look at the case when one or both groups have a sample size less than 30.

5.3 Formula #2: One or Both Groups Have a Sample Size Less Than 30

Suppose that you wanted to see if there was a geographic variation in body length (in mm) of queen honeybees in two different regions of the world (REGION A and REGION B). Suppose, further, that you have been asked to analyze the data from this study and to compare the body lengths of the two regions using the two-group t-test for independent samples. You decide to try out your new Excel skills on a small sample of queen honeybees from each region on the hypothetical data given in Fig. 5.12:

BODY LENGTH (in mm) OF QUEEN HONEYBEES IN TWO REGIONS

REGION A	REGION B
22.56	18.22
20.21	19.21
21.83	19.87
19.02	18.34
20.43	18.65
22.72	19.02
20.65	18.25
18.21	20.73
22.52	19.52
20.23	18.17
	18.75
	18.05

Fig. 5.12 Worksheet Data for Body Length of Queen Honeybees (Practical Example)

Let's call REGION A as Group 1, and REGION B as Group 2.

Null hypothesis: $\mu_1 = \mu_2$
Research hypothesis: $\mu_1 \neq \mu_2$

Note: Since both groups have a sample size less than 30, you need to use Formula #2 in the following steps:

Create an Excel spreadsheet, and enter the following information:

A3: BODY LENGTH (in mm) OF QUEEN HONEYBEES IN TWO REGIONS
B5: REGION A
C5: REGION B
B6: 22.56
B15: 20.23
C6: 18.22
C17: 18.05

Now, enter the other figures into this table. Be sure to double-check all of your figures. If you have only one incorrect figure, you will not be able to obtain the correct answer to this problem.

Now, widen columns B and C so that all of the information fits inside the cells. To do this, click on both letters B and C at the top of these columns on your spreadsheet to highlight all of the cells in columns B and C. Then, move the mouse pointer to the right end of the B cell until you get a "cross" sign; then, click on this cross sign and drag the sign to the right until you can read all of the words on your screen. Then, stop clicking! Both Column B and Column C should now be the same width.

Then, center all information in the table except the top title by using the following steps:

Left-click your mouse and highlight cells B5:C17. Then, click on the bottom line, second from the left icon, under "Alignment" at the top-center of Home. All of the information in the table should now be in the center of each cell.

E6: Null hypothesis:
E8: Research hypothesis:
E11: Group
E12: 1 Region A
E13: 2 Region B
F11: n
G11: Mean
H11: STDEV

Your spreadsheet should now look like Fig. 5.13.

BODY LENGTH (in mm) OF QUEEN HONEYBEES IN TWO REGIONS

REGION A	REGION B				
22.56	18.22	Null hypothesis:			
20.21	19.21				
21.83	19.87	Research hypothesis:			
19.02	18.34				
20.43	18.65				
22.72	19.02	Group	n	Mean	STDEV
20.65	18.25	1 Region A			
18.21	20.73	2 Region B			
22.52	19.52				
20.23	18.17				
	18.75				
	18.05				

Fig. 5.13 Queen Honeybees Body Length Worksheet Data for Hypothesis Testing

Now you need to use your Excel skills from Chap. 1 to fill in the sample sizes (n), the Means, and the Standard Deviations (STDEV) in the Table in cells F12:H13. Be sure to double-check your work or you will not be able to obtain the correct answer to this problem if you have only one incorrect figure!

Since both groups have a sample size less than 30, you need to use Formula #2 for the t-test for the difference of the means of two independent samples.

Formula #2 for the two-group t-test is the following:

$$t = \frac{\overline{X}_1 - \overline{X}_2}{S_{\overline{X}_1 - \overline{X}_2}} \tag{5.1}$$

$$\text{where } S_{\overline{X}_1 - \overline{X}_2} = \sqrt{\frac{(n_1 - 1)S_1^2 + (n_2 - 1)S_2^2}{n_1 + n_2 - 2}\left(\frac{1}{n_1} + \frac{1}{n_2}\right)} \tag{5.5}$$

$$\text{and where degrees of freedom} = df = n_1 + n_2 - 2 \tag{5.6}$$

This formula is complicated, and so it will reduce your chance of making a mistake in writing it if you "break it down into pieces" instead of trying to write the formula as one cell entry.

Now, enter these words on your spreadsheet:

E16: (n1 – 1) × STDEV1 squared
E19: (n2 – 1) × STDEV2 squared
E22: $n_1 + n_2 - 2$
E25: $1/n_1 + 1/n_2$
E28: s.e.
E31: critical t
E34: t-test
B37: Result:
B40: Conclusion: (see Fig. 5.14)

BODY LENGTH (in mm) OF QUEEN HONEYBEES IN TWO REGIONS							
	REGION A	REGION B					
	22.56	18.22		Null hypothesis:			
	20.21	19.21					
	21.83	19.87		Research hypothesis:			
	19.02	18.34					
	20.43	18.65					
	22.72	19.02		Group	n	Mean	STDEV
	20.65	18.25		1 Region A	10	20.84	1.55
	18.21	20.73		2 Region B	12	18.90	0.82
	22.52	19.52					
	20.23	18.17					
		18.75		(n1 - 1) x STDEV1 squared			
		18.05					
				(n2 - 1) x STDEV2 squared			
				n1 + n2 - 2			
				1/n1 + 1/n2			
				s.e.			
				critical t			
				t-test			
	Result:						
	Conclusion:						

Fig. 5.14 Queen Honeybees Body Length Formula Labels for the Two-group t-test

You now need to use your Excel skills to compute the values of the above formulas in the following cells:

H16: the result of the formula needed to compute cell E16 (use two decimals)
H19: the result of the formula needed to compute cell E19 (use two decimals)
H22: the result of the formula needed to compute cell E22
H25: the result of the formula needed to compute cell E25 (use two decimals)
H28: =SQRT(((H16 + H19)/H22)*H25)

Note the three open-parentheses after SQRT, and the three closed parentheses on the right side of this formula. You need three open parentheses and three closed parentheses in this formula or the formula will not work correctly.

The above formula gives a standard error of the difference of the means equal to 0.51 (two decimals) in cell H28.

H31: Enter the critical t value from the t-table in Appendix E in this cell using $df = n_1 + n_2 - 2$ to find the critical t value

H34: =(G12 – G13)/H28 (no spaces between symbols)

Note that you need an open-parenthesis *before G12* and a closed-parenthesis *after G13* so that this answer of 1.94 is *THEN* divided by the standard error of the difference of the means of 0.51, to give a t-test value of 3.77. Use two decimal places for the t-test result (see Fig. 5.15).

BODY LENGTH (in mm) OF QUEEN HONEYBEES IN TWO REGIONS

REGION A	REGION E
22.56	18.22
20.21	19.21
21.83	19.87
19.02	18.34
20.43	18.65
22.72	19.02
20.65	18.25
18.21	20.73
22.52	19.52
20.23	18.17
	18.75
	18.05

Null hypothesis:

Research hypothesis:

Group	n	Mean	STDEV
1 Region A	10	20.84	1.55
2 Region B	12	18.90	0.82

(n1 - 1) x STDEV1 squared	21.50
(n2 - 1) x STDEV2 squared	7.32
n1 + n2 - 2	20
1/n1 + 1/n2	0.18
s.e.	0.51
critical t	2.086
t-test	3.77

Result:

Conclusion:

Fig. 5.15 Queen Honeybees Body Length Two-group t-test Formula Results

Now write the following sentence in C37 to C38 to summarize the *result* of the study:

C37: Since the absolute value of 3.77 is greater than 2.086, we reject the null
C38: hypothesis and accept the research hypothesis.

Finally, write the following sentence in C40 to C41 to summarize the *conclusion* of the study:

C40: The body length of queen honeybees in Region A was significantly longer than
C41: the body length of queen honeybees in Region B (20.84 vs. 18.90).

Save your file as: Honey14

Print the final spreadsheet so that it fits onto one page.
Write the null hypothesis and the research hypothesis by hand on your printout.
The final spreadsheet appears in Fig. 5.16.

BODY LENGTH (in mm) OF QUEEN HONEYBEES IN TWO REGIONS						
REGION A	REGION B					
22.56	18.22	Null hypothesis:		μ_1	=	μ_2
20.21	19.21					
21.83	19.87	Research hypothesis:		μ_1	≠	μ_2
19.02	18.34					
20.43	18.65					
22.72	19.02	Group	n	Mean	STDEV	
20.65	18.25	1 Region A	10	20.84	1.55	
18.21	20.73	2 Region B	12	18.90	0.82	
22.52	19.52					
20.23	18.17					
	18.75	(n1 - 1) x STDEV1 squared			21.50	
	18.05					
		(n2 - 1) x STDEV2 squared			7.32	
		n1 + n2 - 2			20	
		1/n1 + 1/n2			0.18	
		s.e.			0.51	
		critical t			2.086	
		t-test			3.77	
Result:	Since the absolute value of 3.77 is greater than 2.086, we reject the null hypothesis and accept the research hypothesis.					
Conclusion:	The body length of queen honeybees in Region A was significantly longer than the body length of queen honeybees in Region B (20.84 vs. 18.90)					

Fig. 5.16 Queen Honeybees Body Length Final Spreadsheet

5.4 End-of-Chapter Practice Problems

1. Suppose that you wanted to compare the wing length (in mm) of a species of adult mosquitoes in the northeast region and the southeast region of the United States. You have obtained the cooperation of other biologists in Vermont and New Hampshire in the northeast region, and Kentucky and South Carolina in the southeast region who have shared their data with you from a previous study. In the North, 124 mosquitoes had wings with a mean length 3.2 mm and a standard deviation of 1.2 mm. In the South, 135 mosquitoes had wings with a mean length of 3.4 mm with a standard deviation of 1.3 mm.

   (a) State the null hypothesis and the research hypothesis on an Excel spreadsheet.
   (b) Find the standard error of the difference between the means using Excel
   (c) Find the critical t value using Appendix E, and enter it on your spreadsheet.
   (d) Perform a t-test on these data using Excel. What is the value of t that you obtain?
   Use three decimal places for all figures in the formula section of your spreadsheet.
   (e) State your result on your spreadsheet.
   (f) State your conclusion in plain English on your spreadsheet.
   (g) Save the file as: Mosquito3

2. Suppose that you wanted to determine the effect of food quality on the development of house finches (*Carpodacus mexicanus*). To test the effect of food quality on house finch development, you randomly assigned finches to an Experimental group and a Control group. The Experimental group received a special diet of high-quality food, and the Control group received its regular diet of food, during a 120-day period. You then measured their mass (in grams). You want to test your Excel skills on the hypothetical data given in Fig. 5.17.

Fig. 5.17 Worksheet Data for Chap. 5: Practice Problem #2

BODY MASS OF HOUSE FINCHES (in g)	
EXPERIMENTAL GROUP	CONTROL GROUP
19.9	19.8
21.6	19.9
20.1	20.1
21.4	20.5
22.1	20.3
21.1	20.8
21.6	21.1
21.4	21.3
21.8	21.6
22.1	21.8
22.2	22.0
22.6	
22.4	

(a) On your Excel spreadsheet, write the null hypothesis and the research hypothesis.
(b) Create a table that summarizes these data on your spreadsheet and use Excel to find the sample sizes, the means, and the standard deviations of the two groups in this table.
(c) Use Excel to find the standard error of the difference of the means.
(d) Use Excel to perform a two-group t-test. What is the value of t that you obtain (use two decimal places)?
(e) On your spreadsheet, type the *critical value of t* using the t-table in Appendix E.
(f) Type your *result* on the test on your spreadsheet.
(g) Type your *conclusion in plain English* on your spreadsheet.
(h) save the file as: FINCH3

3. Suppose that you have been asked to "run the data" at a local university by the Chair of the Biology Department who wants to know if there is a gender difference in cumulative grade-point averages (GPAs) of Biology majors who have completed all of the required courses in their major. The Chair has obtained the cooperation of the Registrar and has promised to keep the GPA information confidential and has obtained the hypothetical data given in Fig. 5.18:

GPA OF BIOLOGY MAJORS WHO HAVE COMPLETED ALL BIOLOGY REQUIRED COURSES	
MEN	WOMEN
2.45	2.83
2.53	2.74
2.64	2.86
2.72	3.32
2.85	3.36
2.96	3.64
3.01	3.56
3.11	3.56
3.24	3.64
3.35	3.37
3.36	3.67
3.38	3.91
3.21	3.92
3.52	3.64
3.64	3.71
3.75	
3.86	

Fig. 5.18 Worksheet Data for Chap. 5: Practice Problem #3

(a) State the null hypothesis and the research hypothesis on an Excel spreadsheet.
(b) Find the standard error of the difference between the means using Excel
(c) Find the critical t value using Appendix E, and enter it on your spreadsheet.
(d) Perform a t-test on these data using Excel. What is the value of t that you obtain?
(e) State your result on your spreadsheet.
(f) State your conclusion in plain English on your spreadsheet.
(g) Save the file as: GPA3

References

Keller, G. Statistics for Management and Economics (8th ed.). Mason, OH: South-Western Cengage Learning, 2009.

Zikmund, W.G. and Babin, B.J. Exploring Marketing Research (10th ed.). Mason, OH: South-Western Cengage Learning, 2010.

Chapter 6
Correlation and Simple Linear Regression

There are many different types of "correlation coefficients," but the one we will use in this book is the Pearson product-moment correlation which we will call: *r*.

6.1 What Is a "Correlation?"

Basically, a correlation is a number between −1 and +1 that summarizes the relationship between two variables, which we will call X and Y.

A correlation can be either positive or negative. *A positive correlation means that as X increases, Y increases. A negative correlation means that as X increases, Y decreases.* In statistics books, this part of the relationship is called the *direction* of the relationship (i.e., it is either positive or negative).

The correlation also tells us the *magnitude* of the relationship between X and Y. As the correlation approaches closer to +1, we say that the relationship is *strong and positive*.

As the correlation approaches closer to −1, we say that the relationship is *strong and negative*.

A zero correlation means that there is no relationship between X and Y. This means that neither X nor Y can be used as a predictor of the other.

A good way to understand what a correlation means is to see a "picture" of the scatterplot of points produced in a chart by the data points. Let's suppose that you want to know if variable X can be used to predict variable Y. We will place *the predictor variable X on the x-axis* (the horizontal axis of a chart) and *the criterion variable Y on the y-axis* (the vertical axis of a chart). Suppose, further, that you have collected data given in the scatterplots below (see Figs. 6.1, 6.2, 6.3, 6.4, 6.5, and 6.6).

Figure 6.1 shows the scatterplot for a perfect positive correlation of $r = +1.0$. This means that you can perfectly predict each y-value from each x-value because the data points move "upward-and-to-the-right" along a perfectly-fitting straight line (see Fig. 6.1).

T.J. Quirk et al., *Excel 2016 for Biological and Life Sciences Statistics*,
Excel for Statistics, DOI 10.1007/978-3-319-39489-3_6

X	Y
1	1
1.5	1.5
2	2
2.5	2.5
3	3
4	4
4.5	4.5
5	5
5.5	5.5
6	6
3.5	3.5

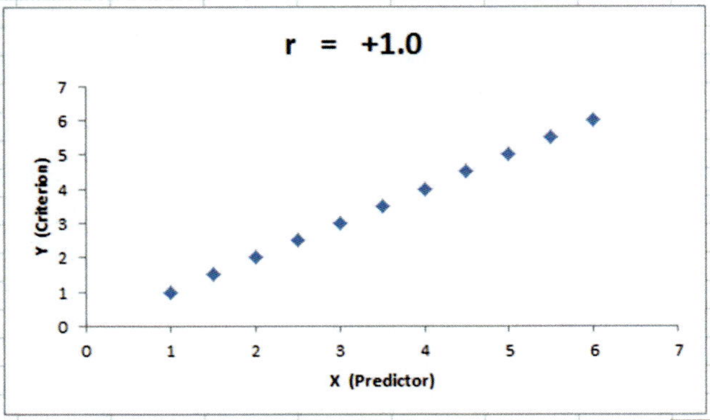

Fig. 6.1 Example of a Scatterplot for a Perfect, Positive Correlation (r = +1.0)

Figure 6.2 shows the scatterplot for a moderately positive correlation of $r = +.53$. This means that each x-value can predict each y-value moderately well because you can draw a picture of a "football" around the outside of the data points that move upward-and-to-the-right, but not along a straight line (see Fig. 6.2).

X	Y
1	2
2	4
3	4.5
4	2.5
5	6
6	5
2.5	2
4	5
5.5	3

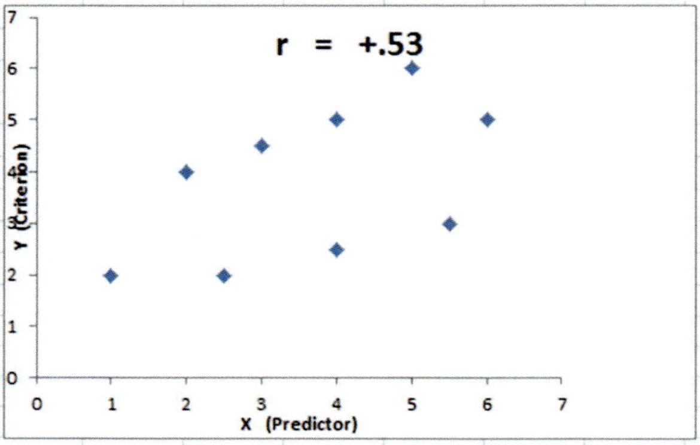

Fig. 6.2 Example of a Scatterplot for a Moderate, Positive Correlation (r = +.53)

Figure 6.3 shows the scatterplot for a low, positive correlation of $r = +.23$. This means that each x-value is a poor predictor of each y-value because the "picture" you could draw around the outside of the data points approaches a circle in shape (see Fig. 6.3).

X	Y
1	2
2	4
3	6
4	1
5	5
6	3
3	1
5	2
4.5	5.5
2	1

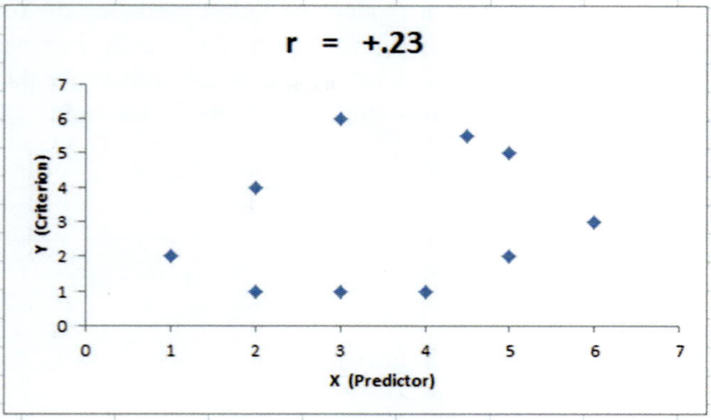

Fig. 6.3 Example of a Scatterplot for a Low, Positive Correlation (r = +.23)

We have not shown a Figure of a zero correlation because it is easy to imagine what it looks like as a scatterplot. A zero correlation of $r = .00$ means that there is no relationship between X and Y and the "picture" drawn around the data points would be a perfect circle in shape, indicating that you cannot use X to predict Y because these two variables are not correlated with one another.

Figure 6.4 shows the scatterplot for a low, negative correlation of $r = -.22$ which means that each X is a poor predictor of Y in an inverse relationship, meaning that as X increases, Y decreases (see Fig. 6.4). In this case, it is a negative correlation because the "football" you could draw around the data points slopes down and to the right.

X	Y
1	6
2	5
2	2
3	1
3	5
5	1
6	3
2.5	2.5
3.5	3.4
5.5	5.4

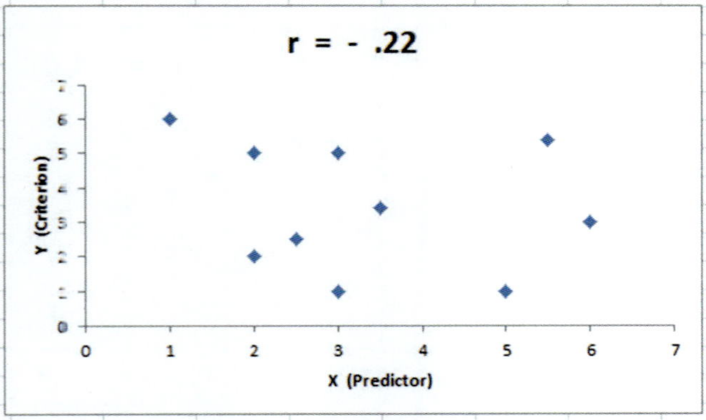

Fig. 6.4 Example of a Scatterplot for a Low, Negative Correlation (r = −.22)

Figure 6.5 shows the scatterplot for a moderate, negative correlation of $r = -.39$ which means that X is a moderately good predictor of Y, although there is an inverse relationship between X and Y (i.e., as X increases, Y decreases; see Fig. 6.5). In this case, it is a negative correlation because the "football" you could draw around the data points slopes down and to the right.

X	Y
1	5
2	6
3	1
4	3
5	4
6	2
1.5	3
5	1
2.5	3
4	6

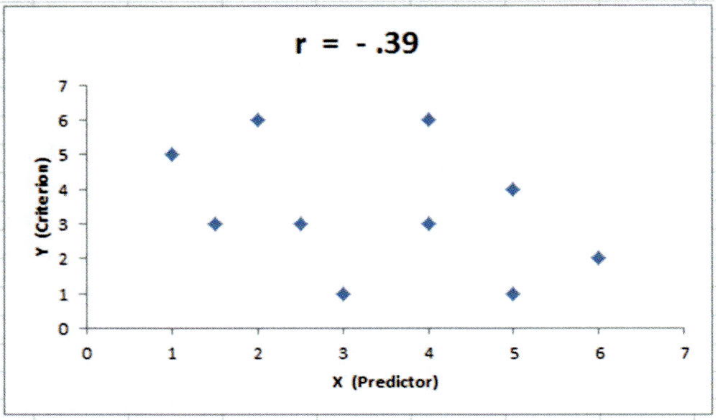

Fig. 6.5 Example of a Scatterplot for a Moderate, Negative Correlation (r = −.39)

Figure 6.6 shows a perfect negative correlation of $r = -1.0$ which means that X is a perfect predictor of Y, although in an inverse relationship such that as X increases, Y decreases. The data points fit perfectly along a downward-sloping straight line (see Fig. 6.6).

X	Y
1	6
1.5	5.5
2	5
2.5	4.5
3	4
4	3
5	2
5.5	1.5
6	1
4.5	2.5
3.5	3.5

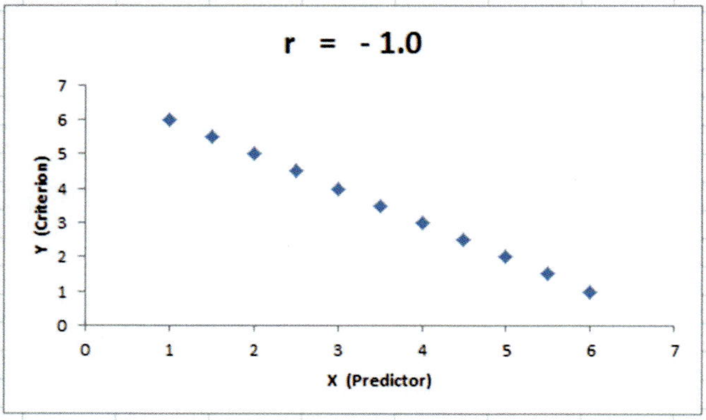

Fig. 6.6 Example of a Scatterplot for a Perfect, Negative Correlation (r = −1.0)

Let's explain the formula for computing the correlation r so that you can understand where the number summarizing the correlation came from.

In order to help you to understand *where* the correlation number that ranges from −1.0 to +1.0 comes from, we will walk you through the steps involved to use the formula as if you were using a pocket calculator. This is the one time in this book that we will ask you to use your pocket calculator to find a correlation, but knowing how the correlation is computed step-by-step will give you the opportunity to understand *how* the formula works in practice.

To do that, let's create a situation in which you need to find the correlation between two variables.

Suppose that you work for a state of the art hydroponic laboratory. You have been asked to conduct a study on the effect of nitrogen on plant growth. You decide to use a pre-determined nitrogen solution and to give varying amounts, in milliliters (ml), of the solution to plants in a laboratory greenhouse. In this example, the x-variable (i.e., the predictor variable) is the amount of nitrogen solution (ml). The y-variable (i.e., the criterion variable) is the amount of plant growth in centimeters (cm). To test your Excel skills, you take a random sample of plants from the laboratory greenhouse. The hypothetical data for eight plants appear in Fig. 6.7. *(Note: We are using only one decimal place for these measurements in this example to simplify the mathematical computations.)*

	A	B	C
1			
2		X	Y
3	Plant	Nitrogen Solution (ml)	Plant Growth (cm)
4	1	2.8	2.9
5	2	2.5	2.8
6	3	3.1	2.8
7	4	3.5	3.2
8	5	2.4	2.6
9	6	2.6	2.3
10	7	2.4	2.1
11	8	3.6	3.2
12			
13	n	8	8
14	MEAN	2.86	2.74
15	STDEV	0.48	0.39
16			

Fig. 6.7 Worksheet Data for Nitrogen Solution and Plant Growth (Practical Example)

Notice also that we have used Excel to find the sample size for both variables, X and Y, and the MEAN and STDEV of both variables. (You can practice your

Excel skills by seeing if you get this same result when you create an Excel spreadsheet for these data.)

Now, let's use the above table to compute the correlation r between nitrogen solution and plant growth using your pocket calculator.

6.1.1 Understanding the Formula for Computing a Correlation

Objective: To understand the formula for computing the correlation r

The formula for computing the correlation r is as follows:

$$r = \frac{\frac{1}{n-1}\Sigma\left(X - \overline{X}\right)\left(Y - \overline{Y}\right)}{S_x\,S_y} \qquad (6.1)$$

This formula looks daunting at first glance, but let's "break it down into its steps" to understand how to compute the correlation r.

6.1.2 Understanding the Nine Steps for Computing a Correlation, r

Objective: To understand the nine steps of computing a correlation r

The nine steps are as follows:

Step	Computation	Result
1	Find the sample size n by noting the number of plants	8
2	Divide the number 1 by the sample size minus 1 (i.e., 1/7)	0.14286
3	*For each Plant*, take its nitrogen solution and subtract the mean nitrogen solution for the 8 plants, and call this $X - \overline{X}$ (For example, for Plant #6, this would be: 2.6 – 2.86) *Note: With your calculator, this difference is –0.26, but when Excel uses 16 decimal places for every computation, this result could be slightly different for each student*	–0.26
4	*For each Plant*, take its growth and subtract the mean Plant growth for the 8 plants, and call this $Y - \overline{Y}$ (For example, for Plant #6, this would be: 2.3 – 2.74)	–0.44
5	Then, *for each Plant*, multiply $\left(X - \overline{X}\right)$ times $\left(Y - \overline{Y}\right)$ (For example, for Plant #6 this would be: (–0.26) × (–0.44)	+0.1144
6	Add the results of $\left(X - \overline{X}\right)$ times $\left(Y - \overline{Y}\right)$ for the 8 plants	+1.09

Steps 1–6 would produce the Excel table given in Fig. 6.8.

	A	B	C	D	E	F
1						
2		X	Y			
3	Plant	Nitrogen Solution (ml)	Plant Growth (cm)	$X-\bar{X}$	$Y-\bar{Y}$	$(X-\bar{X})(Y-\bar{Y})$
4	1	2.8	2.9	-0.06	0.16	-0.01
5	2	2.5	2.8	-0.36	0.06	-0.02
6	3	3.1	2.8	0.24	0.06	0.01
7	4	3.5	3.2	0.64	0.46	0.29
8	5	2.4	2.6	-0.46	-0.14	0.06
9	6	2.6	2.3	-0.26	-0.44	0.11
10	7	2.4	2.1	-0.46	-0.64	0.29
11	8	3.6	3.2	0.74	0.46	0.34
12						-------
13	n	8	8		Total	1.09
14	MEAN	2.86	2.74			
15	STDEV	0.48	0.39			
16						

Fig. 6.8 Worksheet for Computing the Correlation, r

Notice that when Excel multiplies a minus number by a minus number, the result is a plus number (for example for Plant #7: (−0.46) × (−0.64) = +0.29). And when Excel multiplies a minus number by a plus number, the result is a negative number (for example for Plant #1: (−0.06) × (+0.16) = −0.01).

Note: Excel computes all computation to 16 decimal places. So, when you check your work with a calculator, you frequently get a slightly different answer than Excel's answer.

For example, when you compute above:

$$\left(X-\overline{X}\right) \times \left(Y-\overline{Y}\right) \text{ for Plant \#2, your calculator gives:}$$
$$(-0.36) \times (+0.06) = -0.0216 \tag{6.2}$$

As you can see from the table, Excel's answer is −0.02 which is really *more accurate* because Excel uses 16 decimal places for every number, even though only two decimal places are shown in Fig. 6.8.

You should also note that when you do Step 6, you have to be careful to add all of the positive numbers first to get +*1.10* and then add all of the negative numbers second to get −*0.03*, so that when you subtract these two numbers you get +*1.07* as your answer to Step 6. When you do these computations using Excel, this total figure will be +*1.09* because Excel carries every number and computation out to 16 decimal places which is much more accurate than your calculator.

Step		
7	Multiply the answer for step 2 above by the answer for step 6 (0.14286 × 1.09)	0.1557
8	Multiply the STDEV of X times the STDEV of Y (0.48 × 0.39)	0.1872
9	Finally, divide the answer from step 7 by the answer from step 8 (0.1557 divided by 0.1872)	+0.83

This number of *0.83* is the correlation between Nitrogen Solution (X) and Plant Growth (Y) for these 8 plants. The number *+0.83* means that there is a strong, positive correlation between these two variables. That is, as the amount of nitrogen solution increases, plant growth increases. For a more detailed discussion of correlation, see Zikmund and Babin (2010) and McCleery et al. (2007).

You could also use the results of the above table in the formula for computing the correlation r in the following way:

$$\text{correlation r} = \left[(1/(\text{n}-1)) \times \sum \left(X-\overline{X}\right)\left(Y-\overline{Y}\right)\right] / (\text{STDEV}_\text{x} \times \text{STDEV}_\text{y})$$

$$\text{correlation r} = [(1/7) \times 1.09]/[(.48) \times (.39)]$$

$$\text{correlation} = \text{r} = 0.83$$

When you use Excel for these computations, you obtain a slightly different correlation of +0.82 because Excel uses 16 decimal places for all numbers and computations and is, therefore, more accurate than your calculator.

Now, let's discuss how you can use Excel to find the correlation between two variables in a much simpler, and much faster, fashion than using your calculator.

6.2 Using Excel to Compute a Correlation Between Two Variables

Objective: To use Excel to find the correlation between two variables

Suppose that an aquatic biologist in the state of Wisconsin in the U.S.A. has asked you to find the relationship between the weight in grams (g) of a specific species of female fish and the number of eggs produced by that fish at the end of its pregnancy. The aquatic biologist is studying female Walleye fish (*Sander vitreus*) which is a popular fish in the United States and Canada.

To test your Excel skills, you take a random sample of fish near the end of their pregnancy and weigh them (to the nearest gram) and then count the number of eggs they produced (in thousands). The hypothetical data appear in Fig. 6.9.

FEMALE WALLEYE FISH IN WISCONSIN

Is there a relationship between weight and the number of eggs produced?

Weight in grams (g)	Eggs produced (thousands)
1245	58
1467	33
2014	60
2356	65
2478	50
3021	90
3298	84
3588	87
3902	95
3854	85
3667	80
4589	94

Fig. 6.9 Worksheet Data for Weight and Eggs Produced (Practical Example)

Important note: Note that the eggs produced have been rounded off to thousands, so that a fish that produces 65,000 eggs would be recorded as 65.

You want to determine if there is a *relationship* between the weight and eggs produced, and you decide to use a correlation to determine this relationship. Let's call the weight scores the predictor, X, and eggs produced, the criterion, Y.

Create an Excel spreadsheet with the following information:

A3: FEMALE WALLEYE FISH IN WISCONSIN
B5: Is there a relationship between weight and the number of eggs produced?
B7: Weight in grams (g)
C7: Eggs produced (thousands)
B8: 1245
C8: 58

Next, change the width of Columns B and C so that the information fits inside the cells.

Now, complete the remaining figures in the table given above so that B19 is 4589 and C19 is 94. (Be sure to double-check your figures to make sure that they are correct!) Then, center the information in all of these cells.

A21: n
A22: mean
A23: stdev

Next, define the "name" to the range of data from B8:B19 as: weight

We discussed earlier in this book (see Sect. 1.4.4) how to "name a range of data," but here is a reminder of how to do that:

To give a "name" to a range of data:

Click on the top number in the range of data and drag the mouse down to the bottom number of the range.

For example, to give the name: "weight" to the cells: B8:B19, click on B8, and drag the pointer down to B19 so that the cells B8:B19 are highlighted on your computer screen. Then, click on:

Formulas
Define name (top center of your screen)
weight (in the Name box; see Fig. 6.10)

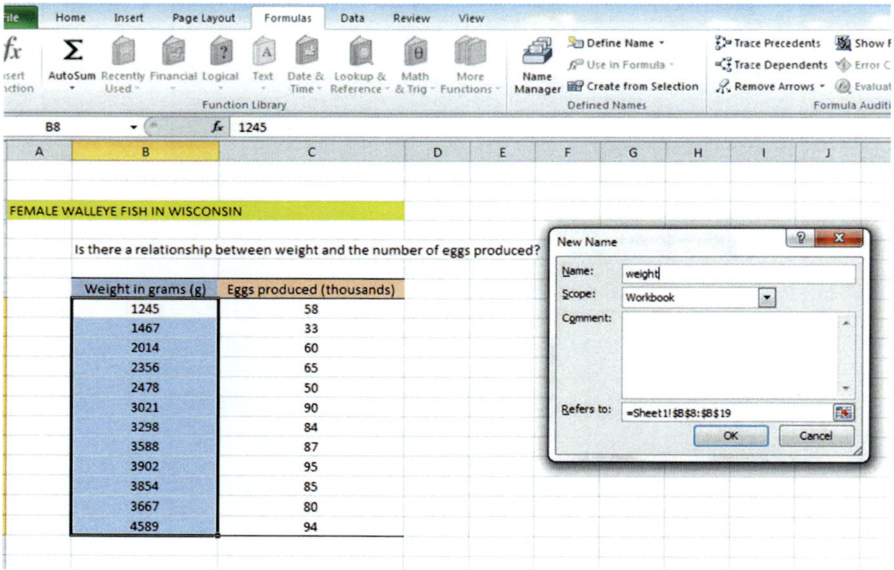

Fig. 6.10 Dialogue Box for Naming a Range of Data as: "weight"

OK

Now, repeat these steps to give the name: *eggs* to C8:C19

Finally, click on any blank cell on your spreadsheet to "deselect" cells C8:C19 on your computer screen.

Now, complete the data for these sample sizes, means, and standard deviations in columns B and C so that B22 is 2956.58, and C23 is 19.79 (use two decimals for the means and standard deviations; see Fig. 6.11)

FEMALE WALLEYE FISH IN WISCONSIN		
	Is there a relationship between weight and the number of eggs produced?	
	Weight in grams (g)	Eggs produced (thousands)
	1245	58
	1467	33
	2014	60
	2356	65
	2478	50
	3021	90
	3298	84
	3588	87
	3902	95
	3854	85
	3667	80
	4589	94
n	12	12
mean	2956.58	73.42
stdev	1045.10	19.79

Fig. 6.11 Example of Using Excel to Find the Sample Size, Mean, and STDEV

Objective: Find the correlation between weight and eggs produced

B25: correlation
C25: =correl(weight,eggs); see Fig. 6.12

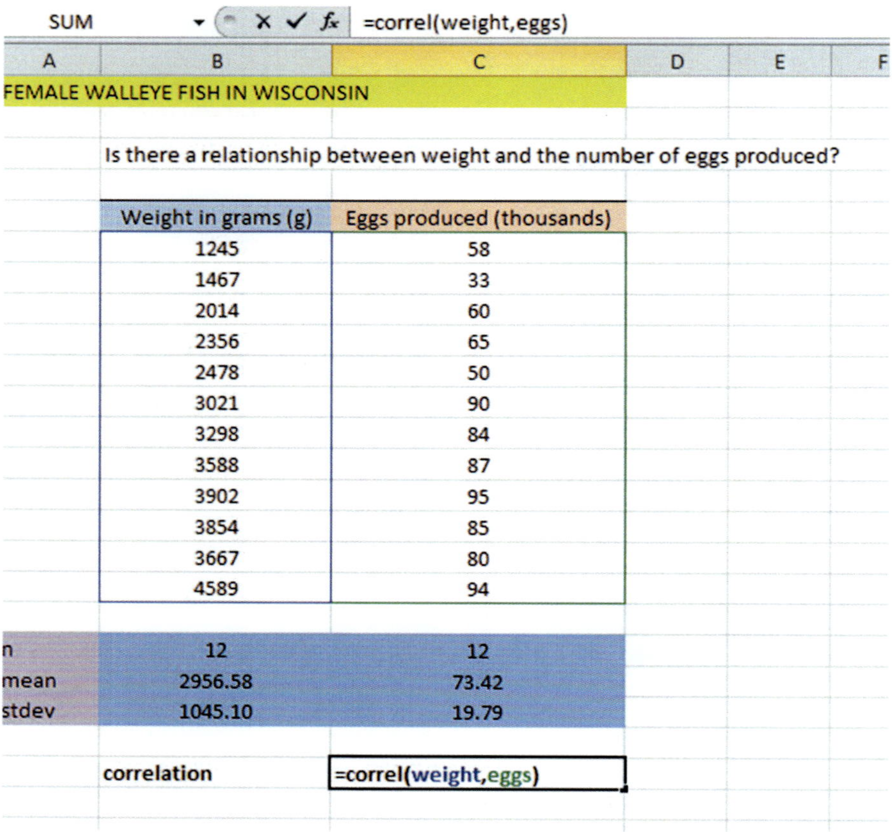

SUM =correl(weight,eggs)

FEMALE WALLEYE FISH IN WISCONSIN

Is there a relationship between weight and the number of eggs produced?

	Weight in grams (g)	Eggs produced (thousands)
	1245	58
	1467	33
	2014	60
	2356	65
	2478	50
	3021	90
	3298	84
	3588	87
	3902	95
	3854	85
	3667	80
	4589	94
n	12	12
mean	2956.58	73.42
stdev	1045.10	19.79
	correlation	=correl(weight,eggs)

Fig. 6.12 Example of Using Excel's =correl Function to Compute the Correlation Coefficient

Hit the Enter key to compute the correlation

C25: format this cell to two decimals

Note that the equal sign in =correl(weight,eggs) in C25 tells Excel that you are going to use a formula in this cell.

The correlation between weight (X) and eggs produced (Y) is +*.87*, a very strong positive correlation. This means that you have evidence that there is a strong relationship between these two variables. In effect, the higher the weight, the more eggs produced by the fish.

Save this file as: eggs33

The final spreadsheet appears in Fig. 6.13.

FEMALE WALLEYE FISH IN WISCONSIN		
	Is there a relationship between weight and the number of eggs produced?	
	Weight in grams (g)	Eggs produced (thousands)
	1245	58
	1467	33
	2014	60
	2356	65
	2478	50
	3021	90
	3298	84
	3588	87
	3902	95
	3854	85
	3667	80
	4589	94
n	12	12
mean	2956.58	73.42
stdev	1045.10	19.79
	correlation	0.87

Fig. 6.13 Final Result of Using the =correl Function to Compute the Correlation Coefficient

6.3 Creating a Chart and Drawing the Regression Line onto the Chart

This section deals with the concept of "linear regression." Technically, the use of a simple linear regression model (i.e., the word "simple" means that only one predictor, X, is used to predict the criterion, Y) requires that the data meet the following four assumptions if that statistical model is to be used:

1. The underlying relationship between the two variables under study (X and Y) is *linear* in the sense that a straight line, and not a curved line, can fit among the data points on the chart.
2. The errors of measurement are independent of each other (e.g. the errors from a specific time period are sometimes correlated with the errors in a previous time period).
3. The errors fit a normal distribution of Y-values at each of the X-values.
4. The variance of the errors is the same for all X-values (i.e., the variability of the Y-values is the same for both low and high values of X).

A detailed explanation of these assumptions is beyond the scope of this book, but the interested reader can find a detailed discussion of these assumptions in Levine et al. (2011, pp. 529–530).

Now, let's create a chart summarizing these data.

Important note: Whenever you are preparing a chart, we strongly recommend that you put the predictor variable (X) on the left, and the criterion variable (Y) on the right in your Excel spreadsheet, so that you do not get these variables backwards in your Excel steps and make a mess of the problem in your computations. If you do this as a habit, you will save yourself a lot of grief whenever you do a problem involving correlation and simple linear regression using Excel.

You need to understand that in any chart that has one predictor and a criterion that there are really TWO LINES that can be drawn between the data points:

Important note: (1) *One line uses X as the predictor, and Y as the criterion.*
(2) *A second line uses Y as the predictor, and X as the criterion.*

This means that you have to be very careful to note in your input data the cells that contain X as the predictor, and Y as the criterion. If you get these cells mixed up and reverse them, you will create the wrong line for your data and you will have botched the problem terribly.

This is why WE STRONGLY RECOMMEND IN THIS BOOK that you always put the X data (i.e., the predictor variable) on the LEFT of your table, and the Y data (i.e., the criterion variable) on the RIGHT of your table on your spreadsheet so that you don't get these variables mixed up.

Also note that the correlation, r, will be exactly the same correlation coefficient (that ranges between −1 and +1) no matter which variable you call the predictor variable and which variable you call the criterion variable. The correlation coefficient just summarizes the relationship between two variables, and it doesn't care which one is the predictor and which one is the criterion.

Let's suppose that you would like to use weight as the predictor variable, and that you would like to use it to predict the number of eggs produced in this species of fish. Since the correlation between these two variables is +*.87*, this shows that there is a strong, positive relationship and that weight is a good predictor of eggs produced.

1. Open the file that you saved earlier in this chapter: eggs33

6.3.1 *Using Excel to Create a Chart and the Regression Line Through the Data Points*

Objective: To create a chart and the regression line summarizing the relationship between weight and eggs produced

2. Click and drag the mouse to highlight both columns of numbers (B8:C19), *but do not highlight the labels at the top of Column B and Column C.*

Highlight the data set: B8:C19
Insert (top left of screen)
Highlight: Scatter chart icon (immediately above the word: "Charts" at the top center of your screen)
Click on the down arrow on the right of the chart icon
Highlight the top left chart icon under "scatter" (see Fig. 6.14)

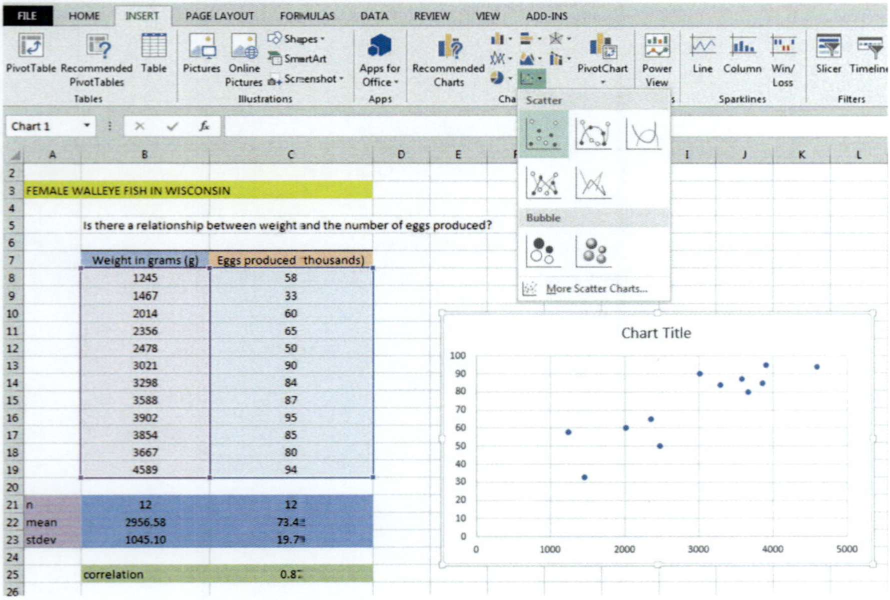

Fig. 6.14 Example of Inserting a Scatter Chart into a Worksheet

Click on the top left chart to select it
Click on the "+ icon" to the right of the chart (CHART ELEMENTS)
Click on the check mark next to "Chart Title" <u>**and also**</u> next to "Gridlines" to remove these check marks (see Fig. 6.15)

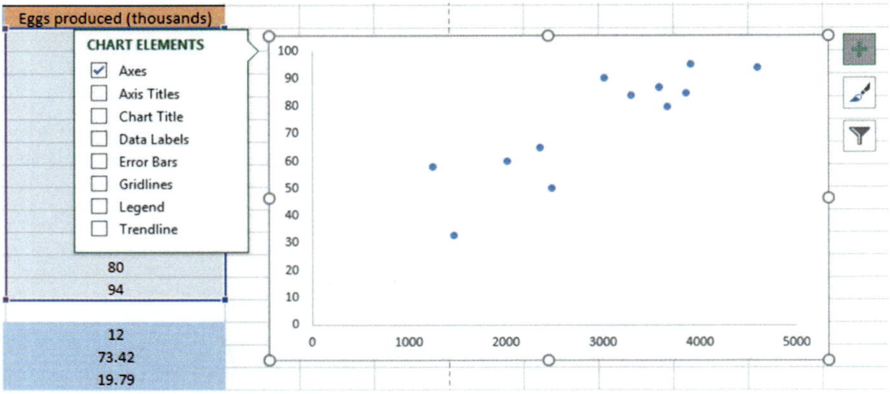

Fig. 6.15 Example of Chart Elements Selected

Click on the box next to: "Chart Title," and then click on the arrow to its right. Then, click on: "Above Chart".

Note that the words: "Chart Title" are now in a box at the top of the chart (see Fig. 6.16).

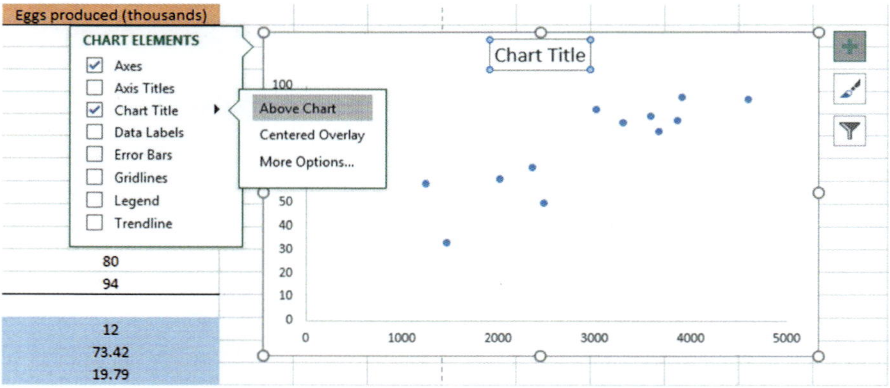

Fig. 6.16 Example of Chart Title Selected

Enter the following Chart Title to the right of $\mathbf{f_x}$ at the top of your screen: RELATIONSHIP BETWEEN WEIGHT AND NUMBER OF EGGS PRODUCED (see Fig. 6.17)

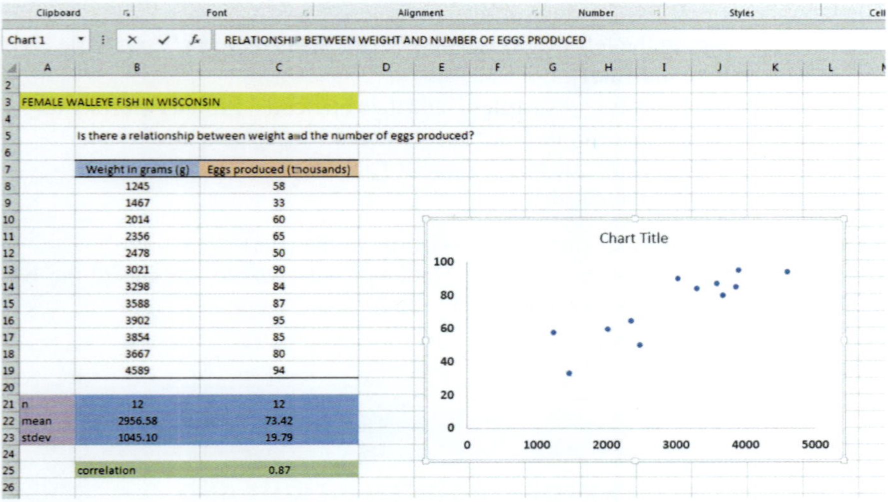

Fig. 6.17 Example of Inserting a Chart Title Above the Chart

Hit the Enter key to place this title above the chart.

Click on *any white space* outside of the top title but inside the chart to "deselect" the box around this chart title (see Fig. 6.18).

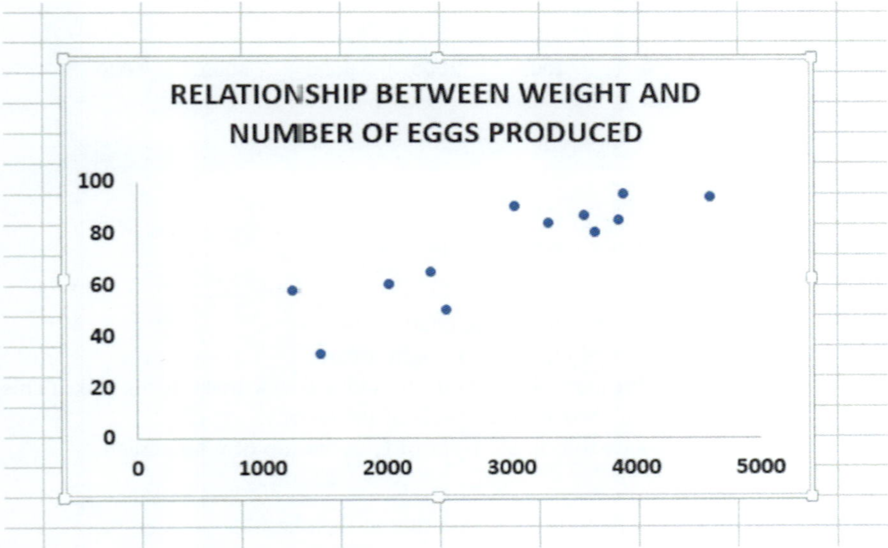

Fig. 6.18 Example of Chart Title Inserted Onto the Chart

Click on the "+ box" to the right of the chart
Add a check mark to the left of "Axis Titles" (This will create an "Axis Title" box on the y-axis of the chart)
Click on the right arrow for "Axis titles," and then click on: "Primary Horizontal" to remove the check mark in its box (this will create the y-axis title)
Enter the following y-axis title to the right of $\mathbf{f_x}$ at the top of your screen
EGGS PRODUCED (thousands)

Then, hit the Enter Key to enter this y-axis title to the chart
Click *inside the chart at the top right corner of the chart* to "deselect" the box around the y-axis title (see Fig. 6.19).

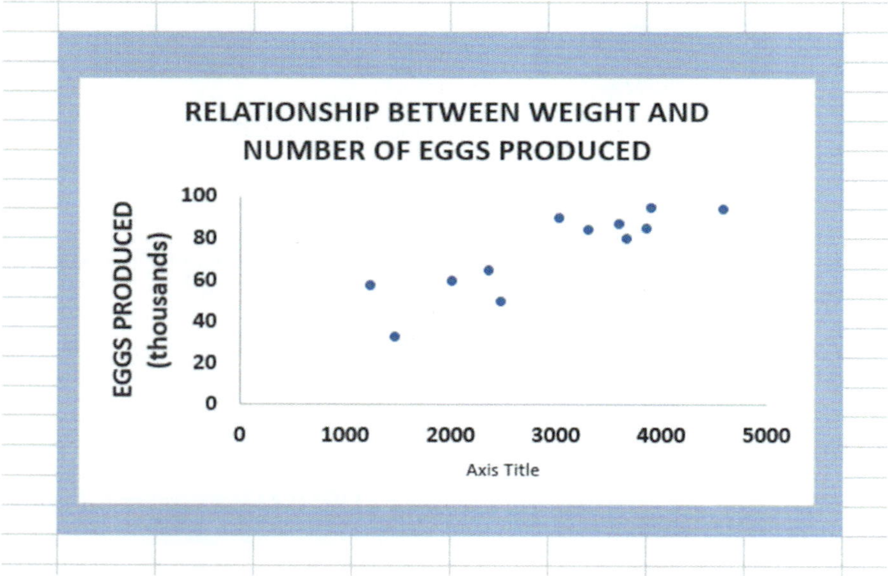

Fig. 6.19 Example of Adding a y-axis Title to the Chart

Click on the "+ box" to the right of the chart
Highlight: "Axis Titles" and click on the right arrow
Click on the words: "Primary Horizontal" to add a check mark to its box. (This creates an "Axis Title" box on the x-axis of the chart.)
Enter the following x-axis title to the right of $\mathbf{f_x}$ at the top of your screen:
WEIGHT (g)

Then, hit the Enter Key to add this x-axis title to the chart.
Click *inside the chart at the top right corner of the chart* to "deselect" the box around the x-axis title (see Fig. 6.20).

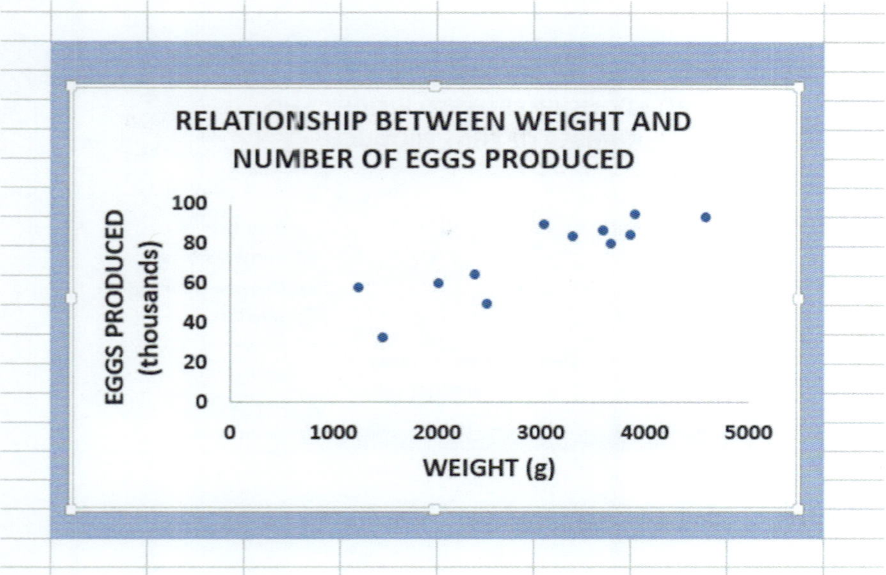

Fig. 6.20 Example of a Chart Title, an x-axis Title, and a y-axis Title

6.3.1.1 Drawing the Regression Line Through the Data Points on the Chart

Objective: To draw the regression line through the data points on the chart

Right-click on any one of the data points inside the chart
Highlight: Add Trendline (see Fig. 6.21).

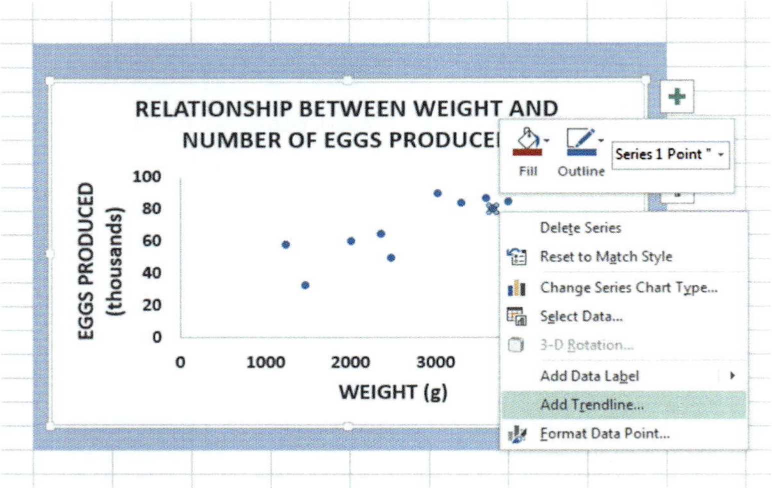

Fig. 6.21 Dialog Box for Adding a Trendline to the Chart

Click on: Add Trendline

Linear (be sure the "linear" button near the top is selected on the "Format Trendline" dialog box) (see Fig. 6.22)

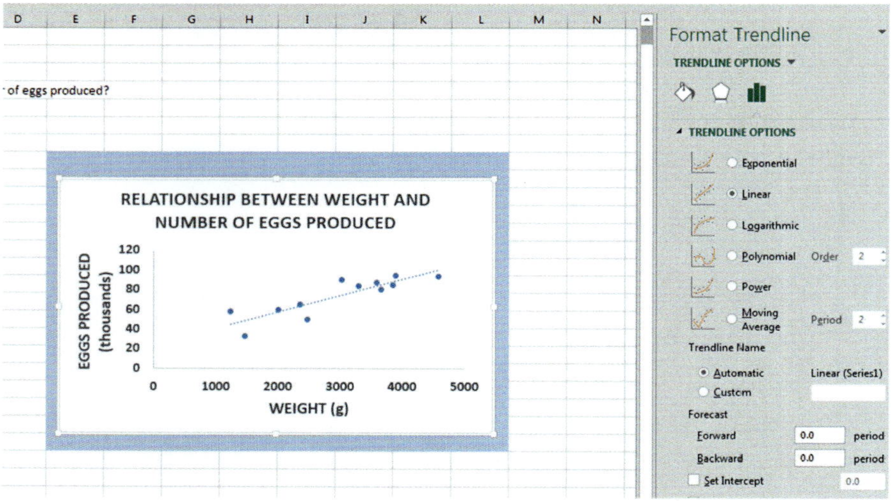

Fig. 6.22 Dialog Box for a Linear Trendline

Click on the "X" at the top right of the "Format Trendline" dialog box to close this dialog box

Click on any blank cell *outside the chart* to "deselect" the chart

Save this file as: eggs34

Your computer monitor screen should look like the spreadsheet in Fig. 6.23.

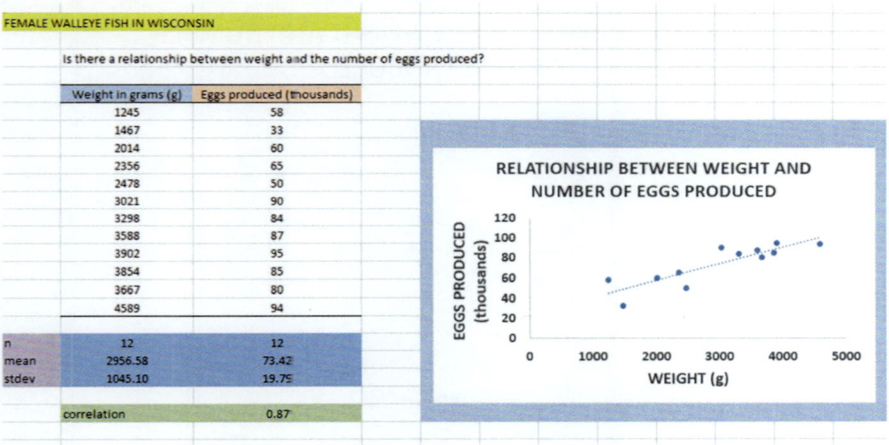

FEMALE WALLEYE FISH IN WISCONSIN

Is there a relationship between weight and the number of eggs produced?

	Weight in grams (g)	Eggs produced (thousands)
	1245	58
	1467	33
	2014	60
	2356	65
	2478	50
	3021	90
	3298	84
	3588	87
	3902	95
	3854	85
	3667	80
	4589	94
n	12	12
mean	2956.58	73.42
stdev	1045.10	19.79
	correlation	0.87

Fig. 6.23 Chart With Trendline Fitted Through the Data Points of the Scatterplot

6.3.1.2 Moving the Chart Below the Table in the Spreadsheet

Objective: To move the chart below the table

Left-click your mouse on *any white space to the right of the top title inside the chart,* keep the left-click down, and drag the chart down and to the left so that the top left corner of the chart is in cell A29, then take your finger off the left-click of the mouse (see Fig. 6.24).

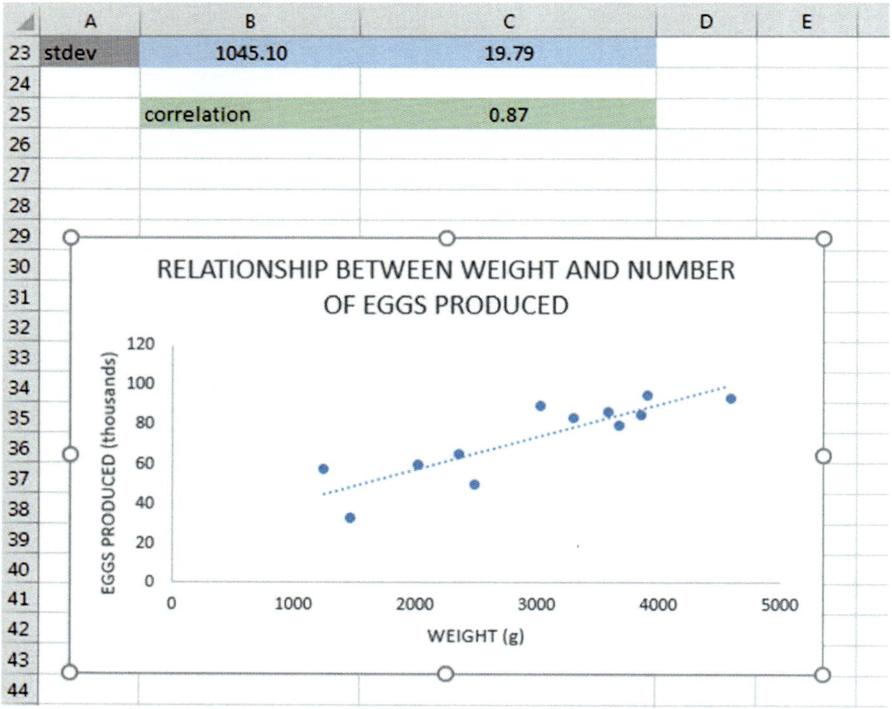

Fig. 6.24 Example of Moving the Chart Below the Table

6.3.1.3 Making the Chart "Longer" So That It Is "Taller"

Objective: To make the chart "longer" so that it is "taller"

Left-click your mouse on the bottom-center of the chart to create a "up-and-down arrow" sign, hold the left-click of the mouse down and drag the bottom of the chart down to row 48 to make the chart longer, and then take your finger off the mouse.

6.3.1.4 Making the Chart "Wider"

Objective: To make the chart "wider"

Put the pointer at the middle of the right-border of the chart to create a "left-to-right arrow" sign, and then left-click your mouse and hold the left-click down while you drag the right-border of the chart to the middle of Column H to make the chart wider.

Now, click on any blank cell outside the chart to "deselect" the chart (see Fig. 6.25).

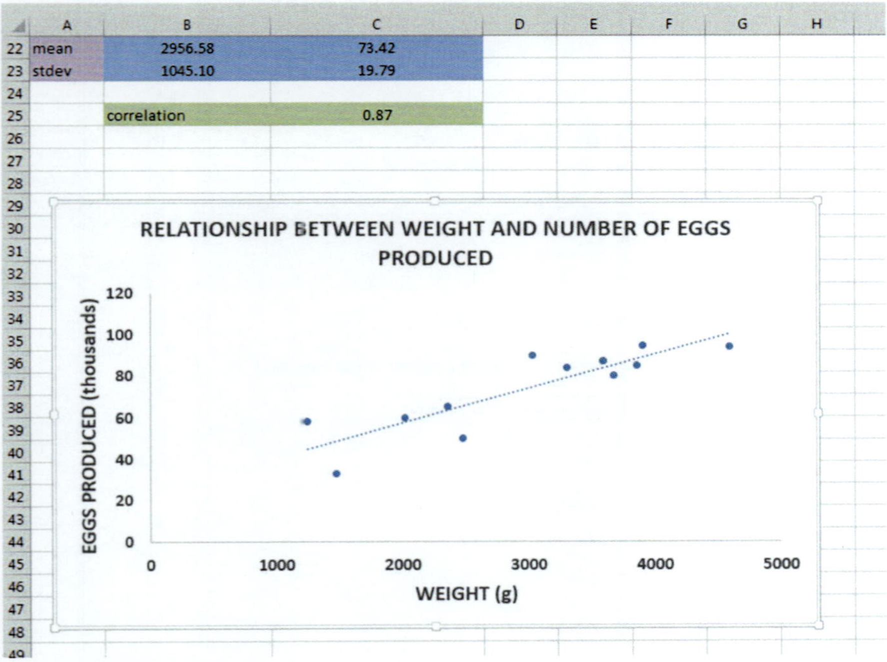

Fig. 6.25 Example of a Chart that is Enlarged to Fit the Ce A29:H48

Now, click on any blank cell outside the chart t 'deselect" the chart.

Save this file as: eggs 35

Note: If you printed this spreadsheet now, is it "t big" to fit onto one page and would "dribble over" onto four pages of pr out because the scale needs to be reduced below 100 % in order for this w sheet to fit onto only one page. You need to complete these next steps belo to print out some, or all, of this spreadsheet.

6.4 Printing a Spreadsheet So That t Table and Chart Fit Onto One Page

Objective: To print the spreadsheet so that t table and the chart fit onto one page

Page Layout (top of screen)

Change the scale at the middle icon near the top of the screen "Scale to Fit" by clicking on the down-arrow until it reads "80 %" so that the table and the chart will fit onto one page on your printout (see Fig. 6.26):

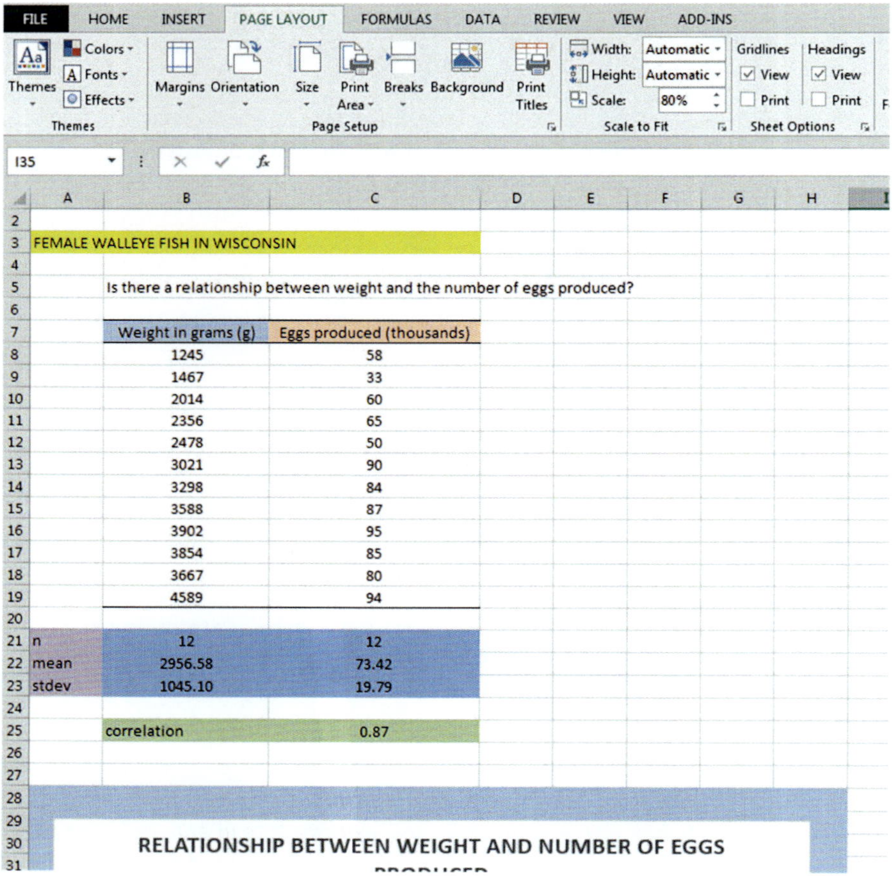

Fig. 6.26 Example of the Page Layout for Reducing the Scale of the Chart to 80 % of Normal Size

File

Print

Print (see Fig. 6.27)

FEMALE WALLEYE FISH IN WISCONSIN

Is there a relationship between weight and the number of eggs produced?

	Weight in grams (g)	Eggs produced (thousands)
	1245	58
	1467	33
	2014	60
	2356	65
	2478	50
	3021	90
	3298	84
	3588	87
	3902	95
	3854	85
	3667	80
	4589	94
n	12	12
mean	2956.58	73.42
stdev	1045.10	19.79
	correlation	0.87

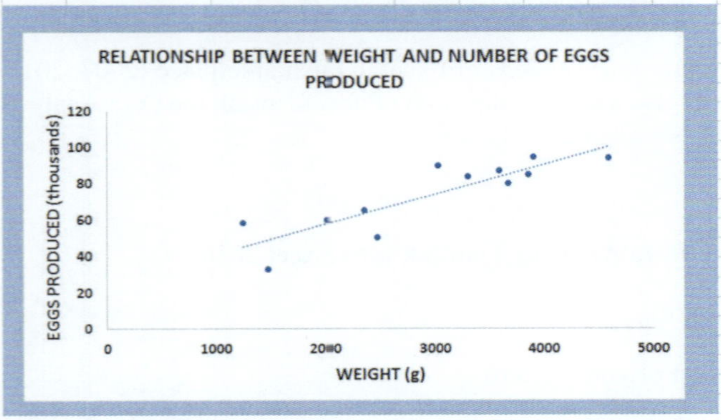

Fig. 6.27 Final Spreadsheet of a Table and a Chart (80 % Scale to Fit Size)

Save your file as: eggs36

6.5 Finding the Regression Equation

The main reason for charting the relationship between X and Y (i.e., weight as X and eggs produced as Y in our example) is to see if there is a strong enough relationship between X and Y so that the regression equation that summarizes this relationship can be used to predict Y for a given value of X.

Since we know that the correlation between the weight of the fish and the number of eggs produced is +.87, this tells us that it makes sense to use weight to predict eggs produced based on past data.

We now need to find that regression equation that is the equation of the "best-fitting straight line" through the data points.

Objective: To find the regression equation summarizing the relationship between X and Y.

In order to find this equation, we need to check to see if your version of Excel contains the "Data Analysis ToolPak" necessary to run a regression analysis.

6.5.1 Installing the Data Analysis ToolPak into Excel

Objective: To install the Data Analysis ToolPak into Excel

Since there are currently four versions of Excel in the marketplace (2007, 2010, 2013, and 2016), we will give a brief explanation of how to install the Data Analysis ToolPak into each of these versions of Excel.

6.5.1.1 Installing the Data Analysis ToolPak into Excel 2016

Open a new Excel spreadsheet

Click on: Data (at the top of your screen)

Look at the top of your monitor screen. Do you see the words: "Data Analysis" at the far right of the screen? If you do, the Data Analysis ToolPak for Excel 2016 was correctly installed when you installed Office 2016, and you should skip ahead to Sect. 6.5.2.

If the words: "Data Analysis" are not at the top right of your monitor screen, then the ToolPak component of Excel 2016 was not installed when you installed Office 2016 onto your computer. If this happens, you need to follow these steps:

File
Options (bottom left of screen)
Note: This creates a dialog box with "Excel Options" at the top left of the box
Add-Ins (on left of screen)
Manage: Excel Add-Ins (at the bottom of the dialog box)
Go (at bottom center of dialog box)
Highlight: Analysis ToolPak (in the Add-Ins dialog box)
Put a check mark to the left of Analysis Toolpak
OK (at the right of this dialog box)
Data

You now should have the words: "Data Analysis" at the top right of your screen to show that this feature has been installed correctly

Note: If these steps do not work, you should try these steps instead:
File/Options (bottom left)/Add-ins/Analysis ToolPak/Go/
click to the left of Analysis ToolPak to add a check mark/OK

If you need help doing this, ask your favorite "computer techie" for help.
You are now ready to skip ahead to Sect. 6.5.2

6.5.1.2 Installing the Data Analysis ToolPak into Excel 2013

Open a new Excel spreadsheet

Click on: Data (at the top of your screen)

Look at the top of your monitor screen. Do you see the words: "Data Analysis" at the far right of the screen? If you do, the Data Analysis ToolPak for Excel 2013 was correctly installed when you installed Office 2013, and you should skip ahead to Sect. 6.5.2.

If the words: "Data Analysis" are not at the top right of your monitor screen, then the ToolPak component of Excel 2013 was not installed when you installed Office 2013 onto your computer. If this happens, you need to follow these steps:

File
Options (bottom left of screen)
Note: This creates a dialog box with "Excel Options" at the top left of the box
Add-Ins (on left of screen)
Manage: Excel Add-Ins (at the bottom of the dialog box)
Go (at bottom center of dialog box)
Highlight: Analysis ToolPak (in the Add-Ins dialog box)
Put a check mark to the left of Analysis Toolpak
OK (at the right of this dialog box)
Data

You now should have the words: "Data Analysis" at the top right of your screen to show that this feature has been installed correctly

If you get a prompt asking you for the "installation CD," put this CD in the CD drive and click on: OK

Note: If these steps do not work, you should try these steps instead:
File/Options (bottom left)/Add-ins/Analysis ToolPak/Go/
click to the left of Analysis ToolPak to add a check mark/OK

If you need help doing this, ask your favorite "computer techie" for help.
You are now ready to skip ahead to Sect. 6.5.2

6.5.1.3 Installing the Data Analysis ToolPak into Excel 2010

Open a new Excel spreadsheet

Click on: Data (at the top of your screen)

Look at the top of your monitor screen. Do you see the words: "Data Analysis" at the far right of the screen? If you do, the Data Analysis ToolPak for Excel 2010 was correctly installed when you installed Office 2010, and you should skip ahead to Sect. 6.5.2.

If the words: "Data Analysis" are not at the top right of your monitor screen, then the ToolPak component of Excel 2010 was not installed when you installed Office 2010 onto your computer. If this happens, you need to follow these steps:

File
Options
Excel options (creates a dialog box)
Add-Ins
Manage: Excel Add-Ins (at the bottom of the dialog box)
Go
Highlight: Analysis ToolPak (in the Add-Ins dialog box)
OK
Data (You now should have the words: "Data Analysis" at the top right of your screen)

If you get a prompt asking you for the "installation CD," put this CD in the CD drive and click on: OK

Note: If these steps do not work, you should try these steps instead:
File/Options (bottom left)/Add-ins/Analysis ToolPak/Go/
click to the left of Analysis ToolPak to add a check mark/OK

If you need help doing this, ask your favorite "computer techie" for help.

You are now ready to skip ahead to Sect. 6.5.2.

6.5.1.4 Installing the Data Analysis ToolPak into Excel 2007

Open a new Excel spreadsheet

Click on: Data (at the top of your screen)

If the words "Data Analysis" do not appear at the top right of your screen, you need to install the Data Analysis ToolPak using the following steps:

Microsoft Office button (top left of your screen)
Excel options (bottom of dialog box)
Add-ins (far left of dialog box)
Go (to create a dialog box for Add-Ins)

Highlight: Analysis ToolPak
OK (If Excel asks you for permission to proceed, click on: Yes)
Data (You should now have the words: "Data Analysis" at the top right of your screen)

If you need help doing this, ask your favorite "computer techie" for help.
You are now ready to skip ahead to Sect. 6.5.2.

6.5.2 *Using Excel to Find the SUMMARY OUTPUT of Regression*

You have now installed *ToolPak*, and you are ready to find the regression equation for the "best-fitting straight line" through the data points by using the following steps:

Open the Excel file: *eggs36* (if it is not already open on your screen)

Note: If this file is already open, and there is a border around the chart, you need to click on any empty cell outside of the chart to deselect the chart.

Now that you have installed *Toolpak*, you are ready to find the regression equation summarizing the relationship between weight and eggs produced in your data set.

Remember that you gave the name: *weight* to the X data (the predictor), and the name: *eggs* to the Y data (the criterion) in a previous section of this chapter (see Sect. 6.2)

Data (top of screen)
Data analysis (far right at top of screen; see Fig. 6.28)

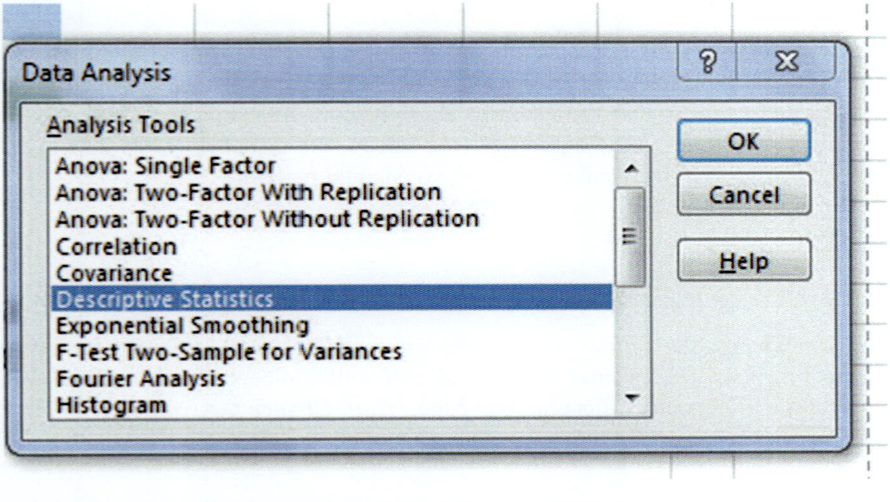

Fig. 6.28 Example of Using the Data/Data Analysis Function of Excel

Scroll down the dialog box using the down arrow and click on: Regression (see Fig. 6.29)

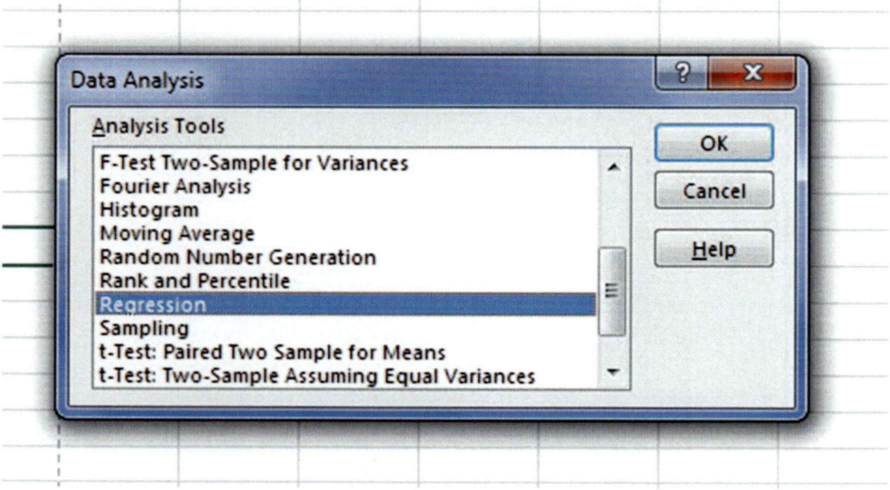

Fig. 6.29 Dialogue Box for Creating the Regression Function in Excel

OK

Input Y Range: eggs
Input X Range: weight

Click on the "button" to the left of Output Range to select this, and enter A50 in the box as the place on your spreadsheet to insert the Regression analysis in cell A50

OK

The *SUMMARY OUTPUT* should now be in cells: A50: I67

Now, make the columns in the Regression Summary Output section of your spreadsheet *wider* so that you can read all of the column headings clearly.

Click on "Home" at the top left of your screen, and then change the data in the following two cells to Number format (two decimal places):

B53
B66

Next, change this cell to four decimal places: B67

Now, change the format for all other numbers that are in decimal format to number format, three decimal places, and center all numbers within their cells.

Print the file so that it fits onto one page. (*Hint: Change the scale under "Page Layout" to 60 % to make it fit.*) Your file should be like the file in Fig. 6.30.

FEMALE WALLEYE FISH IN WISCONSIN

Is there a relationship between weight and the number of eggs produced?

	Weight in grams (g)	Eggs produced (thousands)
	1245	58
	1467	33
	2014	60
	2356	65
	2478	50
	3021	90
	3298	84
	3588	87
	3902	95
	3854	85
	3667	80
	4589	94
n	12	12
mean	2956.58	73.42
stdev	1045.10	19.79
	correlation	0.87

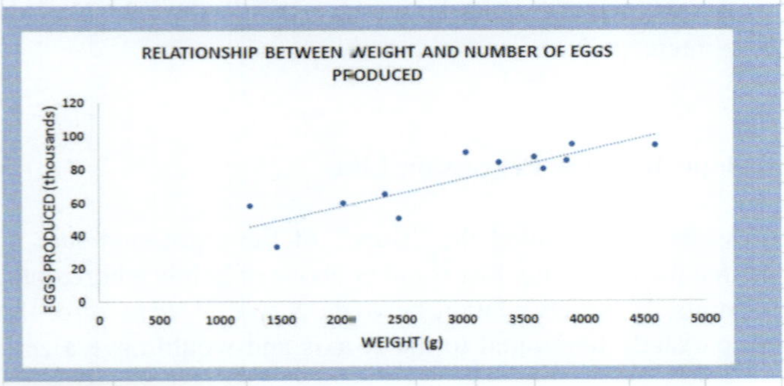

SUMMARY OUTPUT

Regression Statistics	
Multiple R	0.87
R Square	0.756
Adjusted R Square	0.732
Standard Error	10.253
Observations	12

ANOVA

	df	SS	MS	F	Significance F
Regression	1	3257.58[illegible]	3257.589	30.985	0.000
Residual	10	1051.32[illegible]	105.133		
Total	11	4308.91[illegible]			

	Coefficients	Standard Error	t Stat	P-value	Lower 95%	Upper 95%	Lower 95.0%	Upper 95.0%
Intercept	24.73	9.233	2.679	0.023	4.160	45.306	4.160	45.306
X Variable 1	0.0165	0.003	5.566	0.000	0.010	0.023	0.010	0.023

Fig. 6.30 Final Spreadsheet of Correlation and Simple Linear Regression including the SUMMARY OUTPUT for the Data

Save the resulting file as: eggs37

Note the following problem with the summary output.

Whoever wrote the computer program for this version of Excel made a mistake and gave the name: "Multiple R" to cell A53.

This is not correct. Instead, cell A53 should say: "correlation r" since this is the notation that we are using for the correlation between X and Y.

You can now use your printout of the regression analysis to find the regression equation that is the best-fitting straight line through the data points.

But first, let's review some basic terms.

6.5.2.1 Finding the y-intercept, a, of the Regression Line

The point on the y-axis that the regression line would intersect the y-axis if it were extended to reach the y-axis is called the "y-intercept" and *we will use the letter "a" to stand for the y-intercept of the regression line.* The y-intercept on the SUMMARY OUTPUT on the previous page is *24.73 and appears in cell B66.* This means that if you were to draw an imaginary line continuing down the regression line toward the y-axis that this imaginary line would cross the y-axis at 24.73. This is why it is called the "y-intercept."

6.5.2.2 Finding the Slope, b, of the Regression Line

The "tilt" of the regression line is called the "slope" of the regression line. It summarizes to what degree the regression line is either above or below a horizontal line through the data points. If the correlation between X and Y were zero, the regression line would be exactly horizontal to the X-axis and would have a zero slope.

If the correlation between X and Y is positive, the regression line would "slope upward to the right" above the X-axis. Since the regression line in Fig. 6.30 slopes upward to the right, the slope of the regression line is +0.0165 as given in cell *B67. We will use the notation "b" to stand for the slope of the regression line.* (Note that Excel calls the slope of the line: "X Variable 1" in the Excel printout.)

Since the correlation between weight and eggs produced was +*.87*, you can see that the regression line for these data "slopes upward to the right" through the data. Note that the SUMMARY OUTPUT of the regression line in Fig. 6.30 gives a correlation, r, of +*.87* in cell *B53.*

If the correlation between X and Y were negative, the regression line would "slope down to the right" above the X-axis. This would happen whenever the correlation between X and Y is a negative correlation that is between zero and minus one (0 and −1).

6.5.3 Finding the Equation for the Regression Line

To find the regression equation for the straight line that can be used to predict the number of eggs produced from the fish's weight, we only need two numbers in the SUMMARY OUTPUT in Fig. 6.30: *B66 and B67.*

The format for the regression line is : $Y = a + b\ X$ (6.3)

where a = *the y-intercept* (24.73 in our example in cell B66)
and b = *the slope of the line* (+0.0165 in our example in cell B67)

Therefore, the equation for the best-fitting regression line for our example is:

$$\mathrm{Y} = \mathrm{a} + \mathrm{bX}$$

$$\boxed{\mathrm{Y} = 24.73 + 0.0165\mathrm{X}}$$

Remember that Y is the eggs produced that we are trying to predict, using the weight of the fish as the predictor, X.

Let's try an example using this formula to predict the number of eggs produced for a female fish.

6.5.4 *Using the Regression Line to Predict the y-Value for a Given x-Value*

Objective: To find the eggs produced predicted from a fish that weighted 3,000 grams (this is about 6.6 pounds)

Important note: Remember that the weight of the fish is in grams.

Since the weight is 3,000 (i.e., X = 3,000), substituting this number into our regression equation gives:

$$\mathrm{Y} = 24.73 - 0.0165(3000)$$

$$\mathrm{Y} = 24.73 - 49.5$$

$$\mathrm{Y} = 74.23(000) \ \textit{or} \ 74{,}230 \ \textit{eggs}$$

(since the eggs are measured in thousands of eggs)

Important note: If you look at your chart, if you go directly upwards for a weight of 3,000 until you hit the regression line, you see that you hit this line between 70 and 80 on the y-axis to the left when you draw a line horizontal to the x-axis (actually, it is 74.23), the result above for predicting eggs produced from a weight of 3,000 grams.

Now, let's do a second example and predict what the number of eggs produced would be if we used a weight of 3,500 grams.

$Y = 24.73 + 0.0165X$
$Y = 24.73 + 0.0165(3500)$
$Y = 24.73 + 57.75$
$Y = 82.48$ or $82,480$ eggs (since eggs are measured in thousands of eggs)

Important note: If you look at your chart, if you go directly upwards for a weight of 3,500 until you hit the regression line, you see that you hit this line between 80 and 90 on the y-axis to the left (actually it is 82.48), the result above for predicting the number of eggs produced from a fish weighing 3,500 grams.

For a more detailed discussion of regression, see Black (2010) and Gould and Gould (2002).

6.6 Adding the Regression Equation to the Chart

Objective: To Add the Regression Equation to the Chart

If you want to include the regression equation within the chart next to the regression line, you can do that, but a word of caution first.

Throughout this book, we are using the regression equation for one predictor and one criterion to be the following:

$$Y = a + b\ X \tag{6.3}$$

where a = y-intercept and
b = slope of the line

See, for example, the regression equation in Sect. 6.5.3 where the y-intercept was $a = 24.73$ and the slope of the line was $b = +0.0165$ to generate the following regression equation:

$$Y = 24.73 + 0.0165X$$

However, Excel 2016 uses a slightly different regression equation (which is logically identical to the one used in this book) when you add a regression equation to a chart:

$$Y = b\,X + a \tag{6.4}$$

where a = y-intercept and
b = slope of the line

Note that this equation is identical to the one we are using in this book with the terms arranged in a different sequence.

For the example we used in Sect. 6.5.3, Excel 2016 would write the regression equation on the chart as:

$$Y = 0.0165X + 24.73$$

This is the format that will result when you add the regression equation to the chart using Excel 2016 using the following steps:

Open the file: eggs37 (that you saved in Sect. 6.5.2)

Click just *inside* the outer border of the chart in the top right corner to add the "border" around the chart in order to "select the chart" for changes you are about to make

Right-click on any of the data-points in the chart

Highlight: Add Trendline, and click on it to select it

The "Linear button" near the top of the dialog box will be selected (on its left)

Scroll down this dialog box and click on: Display Equation on chart (near the bottom of the dialog box; see Fig. 6.31)

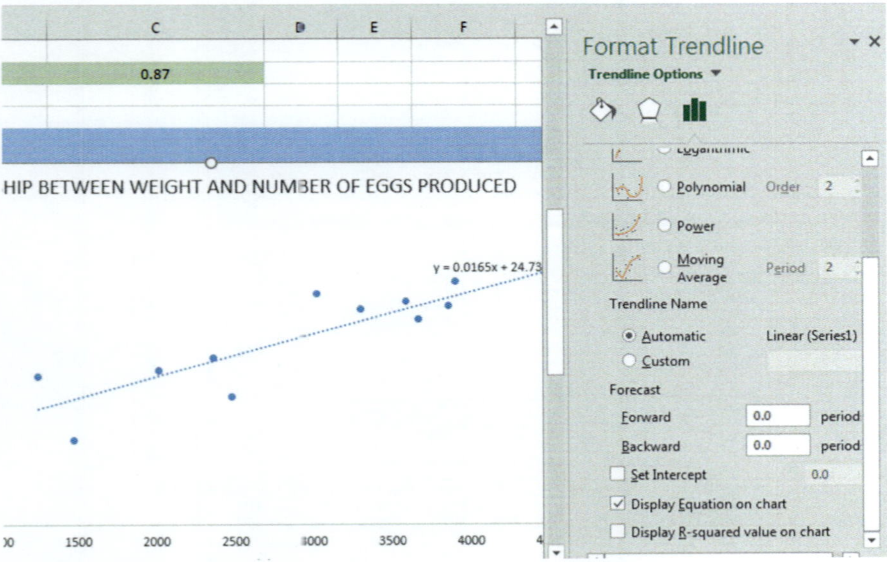

Fig. 6.31 Dialogue Box for Adding the Regression Equation to the Chart Next to the Regression Line on the Chart

Click on the X at the top right of the Format Trendline dialog box to remove this box.

Click on any empty cell outside of the chart to deselect the chart.

Note that the regression equation on the chart is in the following form next to the regression line on the chart (see Fig. 6.32).

FEMALE WALLEYE FISH IN WISCONSIN

Is there a relationship between weight and the number of eggs produced?

Weight in grams (g)	Eggs produced (thousands)
1245	58
1467	33
2014	60
2356	65
2478	50
3021	90
3298	84
3588	87
3902	95
3854	85
3667	80
4589	94

n	12	12
mean	2956.58	73.42
stdev	1045.10	19.79

correlation	0.87

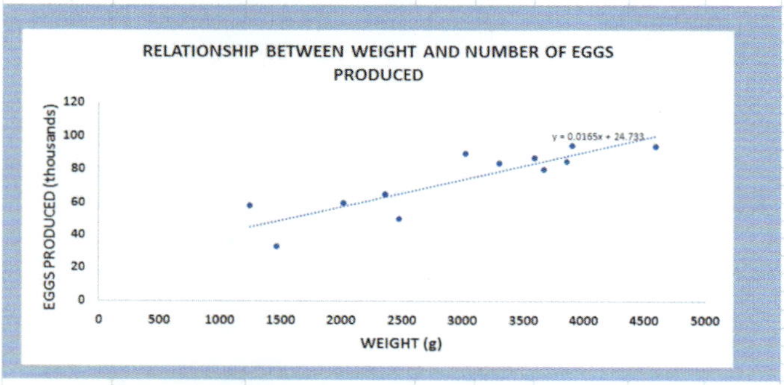

SUMMARY OUTPUT

Regression Statistics	
Multiple R	0.87
R Square	0.756
Adjusted R Square	0.732
Standard Error	10.253
Observations	12

ANOVA

	df	SS	MS	F	Significance F
Regression	1	3257.589	3257.589	30.985	0.000
Residual	10	1051.328	105.133		
Total	11	4308.917			

	Coefficients	Standard Error	t Stat	P-value	Lower 95%	Upper 95%	Lower 95.0%	Upper 95.0%
Intercept	24.73	9.233	2.679	0.023	4.160	45.306	4.160	45.306
X Variable 1	0.0165	0.003	5.566	0.000	0.010	0.023	0.010	0.023

Fig. 6.32 Example of a Chart with the Regression Equation Displayed Next to the Regression Line

$$Y = 0.0165X + 24.73$$

Now, save this file as: eggs38, and print it out so that it fits onto one page.

6.7 How to Recognize Negative Correlations in the SUMMARY OUTPUT Table

Important note: Since Excel does not recognize negative correlations in the SUMMARY OUTPUT results, but treats all correlations as if they were positive correlations (this was a mistake made by the programmer), you need to be careful to note that there may be a negative correlation between X and Y even if the printout says that the correlation is a positive correlation.

You will know that the correlation between X and Y is a negative correlation when these two things occur:

(1) *THE SLOPE, b, IS A NEGATIVE NUMBER. This can only occur when there is a negative correlation.*
(2) *THE CHART CLEARLY SHOWS A DOWNWARD SLOPE IN THE REGRESSION LINE, which can only occur when the correlation between X and Y is negative.*

6.8 Printing Only Part of a Spreadsheet Instead of the Entire Spreadsheet

Objective: To print part of a spreadsheet separately instead of printing the entire spreadsheet

There will be many occasions when your spreadsheet is so large in the number of cells used for your data and charts that you only want to print part of the spreadsheet separately so that the print will not be so small that you cannot read it easily.

We will now explain how to print only part of a spreadsheet onto a separate page by using three examples of how to do that using the file, eggs38, that you created in Sect. 6.6: (1) printing only the table and the chart on a separate page, (2) printing only the chart on a separate page, and (3) printing only the SUMMARY OUTPUT of the regression analysis on a separate page.

Note: If the file: eggs38 is not open on your screen, you need to open it now.

If the border is around the outside of the chart, click on any white space outside of the chart to deselect the chart.

Let's describe how to do these three goals with three separate objectives:

6.8.1 Printing Only the Table and the Chart on a Separate Page

Objective: To print only the table and the chart on a separate page

1. Left-click your mouse starting at the top left of the table *in cell A3* and drag the mouse *down and to the right so that all of the table and all of the chart are highlighted in light blue on your computer screen from cell A3 to cell H48* (the blue cells are called the "selection" cells).
2. File
   Print
   Print Active Sheet (hit the down arrow on the right)
   Print selection
   Print

The resulting printout should contain only the table of the data and the chart resulting from the data.

Then, click on any empty cell in your spreadsheet to deselect the table and chart.

6.8.2 Printing Only the Chart on a Separate Page

Objective: To print only the chart on a separate page

1. Click on any "white space" *just inside the outside border of the chart in the top right corner of the chart* to create the border around all of the borders of the chart in order to "select" the chart.
2. File
   Print
   Print selected chart
   Print selected chart (again)
   Print

The resulting printout should contain only the chart resulting from the data.

Important note: After each time you print a chart by itself on a separate page, you should immediately click on any white space OUTSIDE the chart to remove the border from the chart. When the border is on the borders of the chart, this tells Excel that you want to print only the chart by itself. Do this now!

6.8.3 *Printing Only the SUMMARY OUTPUT of the Regression Analysis on a Separate Page*

Objective: To print only the SUMMARY OUTPUT of the regression analysis on a separate page

1. Left-click your mouse at the cell just above SUMMARY OUTPUT in *cell A50* on the left of your spreadsheet and drag the mouse *down and to the right* until all of the regression output is highlighted in dark blue on your screen from A50 to I67.
2. File
   Print
   Print active sheets (hit the down arrow on the right)
   Print selection
   Print

The resulting printout should contain only the summary output of the regression analysis on a separate page.

Finally, click on any empty cell on the spreadsheet to "deselect" the regression table.

6.9 End-of-Chapter Practice Problems

1. Suppose that you wanted to study the relationship between the number of growth rings and the age of a Ponderosa pine tree (*Pinus ponderosa*). Is there a relationship between these two variables? Let's assume that you can measure the growth rings that correspond to the number of years the tree has been growing.

   You have decided to use the age of the trees (in years) as the predictor, X, and the number of growth rings as the criterion, Y. To test your Excel skills, you have randomly chosen eight trees, and have recorded the hypothetical scores on these variables in Fig. 6.33:

What is the relationship between the age of trees and the number of growth rings on the trees?

AGE (Years)	NUMBER OF GROWTH RINGS
1.0	1.1
5.1	3.5
10.2	8.0
15.3	6.5
20.5	22.0
25.2	17.0
30.1	26.3
34.8	23.6

Fig. 6.33 Worksheet Data for Chap. 6: Practice Problem #1

Create an Excel spreadsheet and enter the data *using AGE as the independent variable (predictor) and NUMBER OF GROWTH RINGS as the dependent variable (criterion).*

Important note: When you are trying to find a correlation between two variables, it is important that you place the predictor, X, ON THE LEFT COLUMN in your Excel spreadsheet, and the criterion, Y, IMMEDIATELY TO THE RIGHT OF THE X COLUMN. You should do this every time that you want to use Excel to find a correlation between two variables to check your thinking so that you do not confuse these two variables with one another.

(a) Create the table using Excel, and then use Excel's =correl function to find the correlation between these two variables, and round off the result to two decimal places. Label the correlation and place it beneath the table.
(b) Create an *XY scatterplot* of these two sets of data such that:

- Top title: RELATIONSHIP BETWEEN AGE AND NUMBER OF GROWTH RINGS IN PONDEROSA PINE TREES
- x-axis title: AGE (years)
- y-axis title: NUMBER OF GROWTH RINGS
- re-size the chart so that it is 8 columns wide and 25 rows long
- move the chart below the table

(c) Create the *least-squares regression line* for these data on the scatterplot and add the regression equation to the chart.
(d) Use Excel to run the regression statistics to find the *equation for the least-squares regression line* for these data and display the results below the chart on your spreadsheet. Use number format (two decimal places) for the correlation and three decimal places for all the other decimal numbers, including the coefficients.

(e) Print just the input data and the chart so that this information fits onto one page. Then, print the regression output table on a separate page so that it fits onto that separate page.
(f) save the file as: GROWTH4
Now, answer these questions using your Excel printout:

1. What is the correlation coefficient *r*?
2. What is the y-intercept?
3. What is the slope of the line?
4. What is the regression equation for these data (use three decimal places for the y-intercept and the slope)?
5. Use the regression equation to predict the NUMBER OF GROWTH RINGS you would expect for a tree that was 20 years old.

2. Suppose that you wanted to study the relationship between the body weight (X) and the gill weight (Y) of the crab, *Pachygrapsus marmoratus*. Body weight was measured in grams (g) while gill weight was measured in milligrams (mg). You are interested in determining if a heavier crab would require larger gills in order to efficiently exchange oxygen and carbon dioxide with its environment.

Create an Excel spreadsheet and enter the data using BODY WEIGHT as the independent (predictor) variable, and GILL WEIGHT as the dependent (criterion) variable. You decide to test your Excel skills on a small sample of crabs using the hypothetical data presented in Fig. 6.34.

BODY WEIGHT VS. GILL WEIGHT IN *Pachygrapsus marmoratus*

BODY WEIGHT IN GRAMS (g)	GILL WEIGHT IN MILLIGRAMS (mg)
3.2	40.2
10.5	90.5
15.6	120.6
17.4	320.7
19.8	240.9
5.3	70.5
13.5	230.7
22.1	370.5
20.4	280.6
18.3	230.4
8.4	165.7

Fig. 6.34 Worksheet Data for Chap. 6: Practice Problem #2

Create an Excel spreadsheet and enter the data using body weight (in grams) as the independent variable (predictor) and gill weight (in milligrams) as the dependent variable (criterion).

(a) create an *XY scatterplot* of these two sets of data such that:

- top title: RELATIONSHIP BETWEEN BODY WEIGHT AND GILL WEIGHT IN *Pachygrapsus marmoratus*
- x-axis title: BODY WEIGHT IN GRAMS (g)
- y-axis title: GILL WEIGHT IN MILLIGRAMS (mg)
- re-size the chart so that it is 7 columns wide and 25 rows long
- move the chart below the table

(b) Create the *least-squares regression line* for these data on the scatterplot.
(c) Use Excel to run the regression statistics to find the *equation for the least-squares regression line* for these data and display the results below the chart on your spreadsheet. Use number format (two decimal places) for the correlation, r, and for both the y-intercept and the slope of the line.
(d) Print the input data and the chart so that this information fits onto one page.
(e) Then, print out the regression output table so that this information fits onto a separate page.
By hand:

(1a) Circle and label the value of the *y-intercept* and the *slope* of the regression line onto that separate page.
(2b) *Read from the graph* the gill weight you would predict for a *body weight of 20 grams* and write your answer in the space immediately below:

____

(f) save the file as: CRAB3
Answer the following questions using your Excel printout:

1. What is the correlation?
2. What is the y-intercept?
3. What is the slope of the line?
4. What is the regression equation for these data (use two decimal places for the y-intercept and the slope)?
5. Use that regression equation to predict the gill weight you would expect for a body weight of 15 grams.

(Note that this correlation is not the multiple correlation as the Excel table indicates, but is merely the correlation r instead.)

Note that you found a positive correlation of +.87 between body weight and gill weight in crabs above. You know that the correlation is a positive correlation for two reasons: (1) the regression line slopes upward and to the right on the chart, signaling a positive correlation, and (2) the slope is +14.58 which also tells you that the correlation is a positive correlation.

But how does Excel treat *negative correlations*?

Important note: Since Excel does not recognize negative correlations in the SUMMARY OUTPUT but treats all correlations as if they were positive correlations, you need to be careful to note when there is a negative correlation between the two variables under study.

You know that the correlation is negative when:

(1) *The slope, b, is a negative number which can only occur when there is a negative correlation.*
(2) *The chart clearly shows a downward slope in the regression line, which can only happen when the correlation is negative.*

3. Suppose that you wanted to study mayflies in lakes in western Montana. Mayflies are common aquatic insects found in rivers, streams, and lakes across the United States. You are trying to study the relationship between the body length and the forewing length of mayflies.

   You want to study the body length (from the head to the start of the tail) in millimeters (mm) and the forewing (front wing) length (in mm). You have decided to use body length as the predictor and forewing length as the criterion. You have collected a small sample of mayflies from various lakes in Montana to test your Excel skills and to make sure that you can do this type of research. The hypothetical data appear in Fig. 6.35:

Is there a relationship between body length and forewing length in mayflies?

Body Length (mm)	Forewing Length (mm)
5.60	4.20
6.20	5.10
5.80	4.60
6.40	5.20
7.10	5.90
7.20	6.00
6.20	5.10
6.50	5.20
6.80	5.30
6.90	5.70
5.60	4.30

Fig. 6.35 Worksheet Data for Chap. 6: Practice Problem #3

Create an Excel spreadsheet and enter the data using body length as the independent variable (predictor) and the forewing length as the dependent variable (criterion). Underneath the table, use Excel's =*correl* function to find the correlation between these two variables. Label the correlation and place it underneath the table; then round off the correlation to two decimal places.

(a) create an *XY scatterplot* of these two sets of data such that:

- top title: RELATIONSHIP BETWEEN BODY LENGTH (mm) AND FOREWING LENGTH (mm) IN MAYFLIES
- x-axis title: Body Length (mm)
- y-axis title: Forewing Length (mm)
- move the chart below the table
- re-size the chart so that it is 8 columns wide and 25 rows long
- move the chart below the table

(b) Create the *least-squares regression line* for these data on the scatterplot, and add the regression equation to the chart.
(c) Use Excel to run the regression statistics to find the *equation for the least-squares regression line* for these data and display the results below the chart on your spreadsheet. Use number format (two decimal places) for the correlation and three decimal places for all other decimal figures, including the coefficients.
(d) Print just the input data and the chart so that this information fits onto one page. Then, print the regression output table on a separate page so that it fits onto that separate page.
(e) save the file as: BODY3
Answer the following questions using your Excel printout:

1. What is the correlation between Body Length and Forewing Length?
2. What is the y-intercept?
3. What is the slope of the line?
4. What is the regression equation?
5. Use the regression equation to predict the Forewing Length you would expect for a mayfly that had a body length of 6.50 mm. Show your work on a separate sheet of paper.

References

Black, K. Business Statistics: For Contemporary Decision Making (6th ed.). Hoboken, NJ: John Wiley & Sons, Inc., 2010.

Gould, J.L. and Gould, G.F. Biostats Basics: A Student Handbook. New York, NY: W.H. Freeman and Company, 2002.

Levine, D.M.. Stephan, D.F., Krehbiel, T.C., and Berenson, M.L. Statistics for Managers Using Microsoft Excel (6th ed.). Boston, MA: Prentice Hall/Pearson, 2011.

McCleery, R., Watt, T. and Hart, T. Introduction to Statistics for Biology (3rd ed.). Boca Raton, FL: Chapman & Hall/CRC, 2007.

Zikmund, W.G. and Babin, B.J. Exploring Marketing Research (10th ed.). Mason, OH: South-Western Cengage Learning, 2010.

Chapter 7
Multiple Correlation and Multiple Regression

There are many times in science when you want to predict a criterion, Y, but you want to find out if you can develop a better prediction model by using *several predictors* in combination (e.g. X_1, X_2, X_3, etc.) instead of a single predictor, X.

The resulting statistical procedure is called "multiple correlation" because it uses two or more predictors in combination to predict Y, instead of a single predictor, X. Each predictor is "weighted" differently based on its separate correlation with Y and its correlation with the other predictors. The job of multiple correlation is to produce a regression equation that will weight each predictor differently and in such a way that the combination of predictors does a better job of predicting Y than any single predictor by itself. We will call the multiple correlation: R_{xy}.

You will recall (see Sect. 6.5.3) that the regression equation that predicts Y when only one predictor, X, is used is:

$$Y = a + bX \tag{7.1}$$

7.1 Multiple Regression Equation

The multiple regression equation follows a similar format and is:

$$Y = a + b_1X_1 + b_2X_2 + b_3X_3 + \textit{etc. depending on the number of predictors used} \tag{7.2}$$

The "weight" given to each predictor in the equation is represented by the letter "b" with a subscript to correspond to the same subscript on the predictors.

You will remember from Chap. 6 that the correlation, r, ranges from −1 to +1. However, the multiple correlation, R_{xy}, only ranges from zero to +1. R_{xy} is never a negative number!

T.J. Quirk et al., *Excel 2016 for Biological and Life Sciences Statistics*, Excel for Statistics, DOI 10.1007/978-3-319-39489-3_7

Important note: In order to do multiple regression, you need to have installed the "Data Analysis ToolPak" that was described in Chap. 6 (see Sect. 6.5.1). If you did not install this, you need to do so now

Let's try a practice problem.

Suppose that you have been asked to analyze some data on fruit production from a mid-size fruit farm in Colorado. The farm operates a laboratory greenhouse where it tests new fruit growth under controlled conditions. The farm is able to control the amount of water (milliliters, ml), the amount of light received (minutes, min), and the amount of fertilizer (microliters, μl) a fruit plant receives. Suppose that the farm wants you to determine the relationship between water, light, and fertilizer in their ability to predict the mass of fruit (grams, g) produced. The farm manager has also asked you to look at the amount of fruit produced in relation to growing conditions.

You have decided to use the three measured growing conditions as the predictors (X_1, X_2, and X_3) and the mass of fruit produced as the criterion, Y. To test your Excel skills, you have randomly selected 11 plants and recorded their growing conditions.

Let's use the following notation:

Y	FRUIT PRODUCED
X_1	WATER
X_2	LIGHT RECEIVED
X_3	FERTILIZER

Suppose, further, that you have collected the following hypothetical data summarizing these scores (see Fig. 7.1):

FRUIT PRODUCED IN RELATION TO GROWING CONDITIONS

Is there a relationship between fruit produced and growing conditions?

FRUIT PRODUCED (g)	WATER (ml)	LIGHT RECEIVED (min)	FERTILIZER (μl)
2.55	250	230	220
3.05	610	240	440
3.55	620	540	530
2.05	420	420	260
2.45	320	520	320
2.95	630	620	620
3.15	650	540	530
3.45	520	580	560
3.30	420	490	630
2.75	330	220	610
3.65	440	570	660

Fig. 7.1 Worksheet Data for Growing Conditions versus Fruit Produced (Practical Example)

Create an Excel spreadsheet for these data using the following cell reference:

A2: FRUIT PRODUCED IN RELATION TO GROWING CONDITIONS
A4: Is there a relationship between fruit produced and growing conditions?
A6: FRUIT PRODUCED (g)
A7: 2.55
B6: WATER (ml)
C6: LIGHT RECEIVED (min)
D6: FERTILIZER (μl)
D17: 660

Next, change the column width to match the above table, and change all FRUIT PRODUCED figures to number format (two decimal places).

Now, fill in the additional data in the chart such that:

A17: 3.65
B17: 440
C17: 570

Then, center all numbers in your table

Important note: Be sure to double-check all of your numbers in your table to be sure that they are correct, or your spreadsheets will be incorrect.

Save this file as: FRUIT3

Before we do the multiple regression analysis, we need to try to make one important point very clear:

Important: When we used one predictor, X, to predict one criterion, Y, we said that you need to make sure that the X variable is ON THE LEFT in your table, and the Y variable is ON THE RIGHT in your table so that you don't get these variables mixed up (see Sect. 6.3)

However, in multiple regression, you need to follow this rule which is exactly the opposite:

When you use several predictors in multiple regression, it is essential that the criterion you are trying to predict, Y, be ON THE FAR LEFT, and all of the predictors are TO THE RIGHT of the criterion, Y, in your table so that you know which variable is the criterion, Y, and which variables are the predictors. If you make this a habit, you will save yourself a lot of grief.

Notice in the table above, that the criterion Y (FRUIT PRODUCED) is on the far left of the table, and the three predictors (WATER, LIGHT RECEIVED, and FERTILIZER) are to the right of the criterion variable. If you follow this rule, you will be less likely to make a mistake in this type of analysis.

7.2 Finding the Multiple Correlation and the Multiple Regression Equation

Objective: To find the multiple correlation and multiple regression equation using Excel.

You do this by the following commands:
Data
Click on: Data Analysis (far right top of screen)
Regression (scroll down to this in the box; see Fig. 7.2)

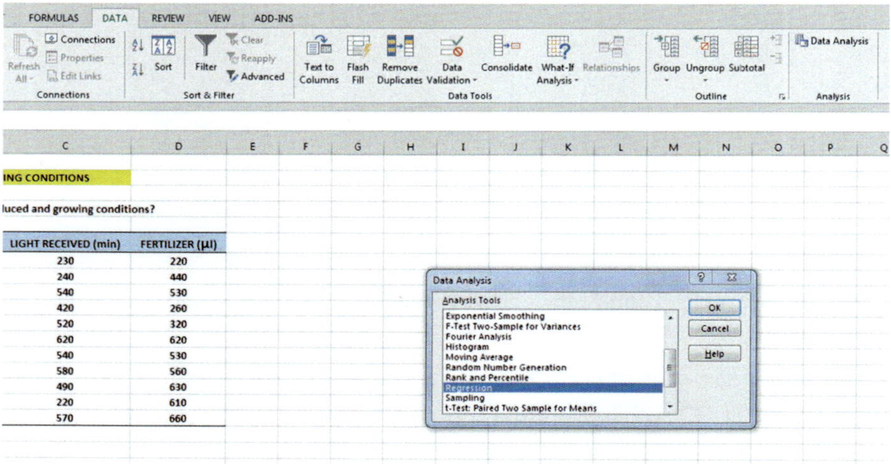

Fig. 7.2 Dialogue Box for Regression Function

OK

Input Y Range: A6:A17
Input X Range: B6:D17

Note that both the input Y Range and the Input X Range above both include the label at the top of the columns.

Click on the Labels box to *add a check mark* to it (because you have included the column labels in row 6)

Output Range (click on the button to its left, and enter): A20 (see Fig. 7.3)

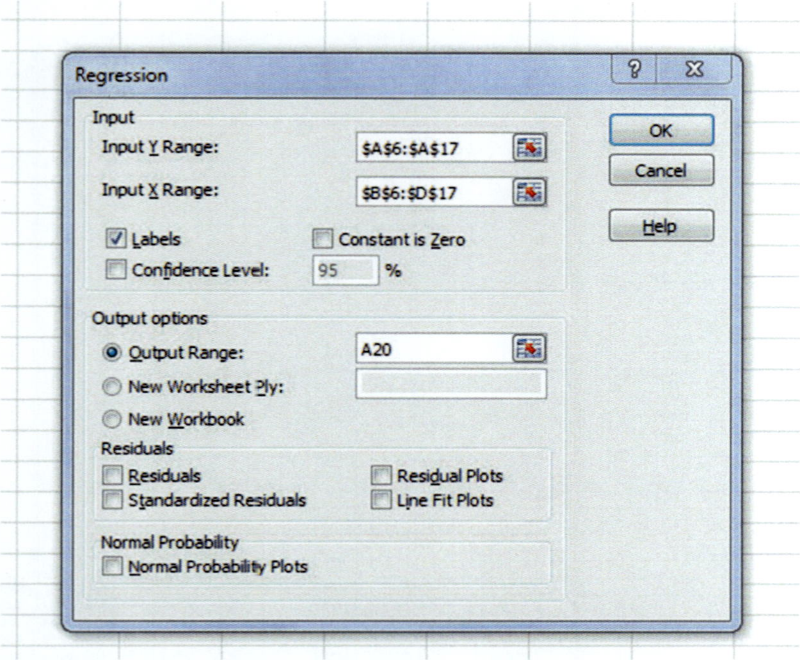

Fig. 7.3 Dialogue Box for Growing Conditions vs. Fruit Produced Data

Important note: Excel automatically assigns a dollar sign $ in front of each column letter and each row number so that you can keep these ranges of data constant for the regression analysis.

OK (see Fig. 7.4 to see the resulting SUMMARY OUTPUT)

FRUIT PRODUCED IN RELATION TO GROWING CONDITIONS

Is there a relationship between fruit produced and growing conditions?

FRUIT PRODUCED (g)	WATER (ml)	LIGHT RECEIVED (min)	FERTILIZER (μl)
2.55	250	230	220
3.05	610	240	440
3.55	620	540	530
2.05	420	420	260
2.45	320	520	320
2.95	630	620	620
3.15	650	540	530
3.45	520	580	560
3.30	420	490	630
2.75	330	220	610
3.65	440	570	660

SUMMARY OUTPUT

Regression Statistics	
Multiple R	0.797651156
R Square	0.636247366
Adjusted R Square	0.48035338
Standard Error	0.361446932
Observations	11

ANOVA

	df	*SS*	*MS*	*F*	*Significance F*
Regression	3	1.599583719	0.533194573	4.081282	0.057174747
Residual	7	0.91450719	0.130643884		
Total	10	2.514090909			

	Coefficients	*Standard Error*	*t Stat*	*P-value*	*Lower 95%*	*Upper 95%*	*Lower 95.0%*	*Upper 95.0%*
Intercept	1.53627108	0.468442063	3.279532734	0.013496	0.428581617	2.643960543	0.428581617	2.643960543
WATER (ml)	0.000642945	0.000963026	0.667629207	0.525762	-0.001634251	0.00292014	-0.001634251	0.00292014
LIGHT RECEIVED (min)	0.000264354	0.000889915	0.297055329	0.775046	-0.00183996	0.002368667	-0.00183996	0.002368667
FERTILIZER (μl)	0.00210733	0.000848684	2.4830572	0.042022	0.000100512	0.004114149	0.000100512	0.004114149

Fig. 7.4 Regression SUMMARY OUTPUT of Growing Conditions vs. Fruit Produced Data

Next, format cell B23 in number format (two decimal places)

Next, format the following four cells in Number format (four decimal places):

B36
B37
B38
B39

Change all other decimal figures to two decimal places, and center all figures within their cells.

Save the file as: FRUIT4

Now, print the file so that it fits onto one page by changing the scale to *55 % size*. The resulting regression analysis is given in Fig. 7.5.

FRUIT PRODUCED IN RELATION TO GROWING CONDITIONS

Is there a relationship between fruit produced and growing conditions?

FRUIT PRODUCED (g)	WATER (ml)	LIGHT RECEIVED (min)	FERTILIZER (μl)
2.55	250	230	220
3.05	610	240	440
3.55	620	540	530
2.05	420	420	260
2.45	320	520	320
2.95	630	620	620
3.15	650	540	530
3.45	520	580	560
3.30	420	490	630
2.75	330	220	610
3.65	440	570	660

SUMMARY OUTPUT

Regression Statistics	
Multiple R	0.80
R Square	0.64
Adjusted R Square	0.48
Standard Error	0.36
Observations	11

ANOVA

	df	*SS*	*MS*	*F*	*Significance F*
Regression	3	1.60	0.53	4.08	0.06
Residual	7	0.91	0.13		
Total	10	2.51			

	Coefficients	*Standard Error*	*t Stat*	*P-value*	*Lower 95%*	*Upper 95%*	*Lower 95.0%*	*Upper 95.0%*
Intercept	1.5363	0.47	3.28	0.01	0.43	2.64	0.43	2.64
WATER (ml)	0.0006	0.00	0.67	0.53	0.00	0.00	0.00	0.00
LIGHT RECEIVED (min)	0.0003	0.00	0.30	0.78	0.00	0.00	0.00	0.00
FERTILIZER (μl)	0.0021	0.00	2.48	0.04	0.00	0.00	0.00	0.00

Fig. 7.5 Final Spreadsheet for Growing Conditions vs. Fruit Produced Regression Analysis

Once you have the SUMMARY OUTPUT, you can determine the multiple correlation and the regression equation that is the best-fit line through the data points using WATER, LIGHT RECEIVED, and FERTILIZER as the three predictors, and FRUIT PRODUCED as the criterion.

Note on the SUMMARY OUTPUT where it says: "Multiple R." This term is correct since this is the term Excel uses for the multiple correlation, which is +0.80. This means, that from these data, that the combination of WATER, LIGHT RECEIVED, and FERTILIZER together form a very strong positive relationship in predicting FRUIT PRODUCED.

To find the regression equation, *notice the coefficients at the bottom of the SUMMARY OUTPUT*:

Intercept: a (this is the y-intercept)	*1.5363*
WATER*: b_1*	*0.0006*
LIGHT RECEIVED: b_2	*0.0003*
FERTILIZER: b_3	0.0021

Since the general form of the multiple regression equation is:

$$Y = a + b_1X_1 + b_2X_2 + b_3X_3 \tag{7.2}$$

we can now write the multiple regression equation for these data:

$$Y = 1.5363 + 0.0006X_1 + 0.0003X_2 + 0.0021X_3$$

7.3 Using the Regression Equation to Predict FRUIT PRODUCED

> Objective: To find the predicted FRUIT PRODUCED using an a WATER Score of 600, a LIGHT RECEIVED Score of 500, and a FERTILIZER Score of 550

Plugging these three numbers into our regression equation gives us:

$$\begin{aligned} Y &= 1.5363 + 0.0006(600) + 0.0003(500) + 0.0021(550) \\ Y &= 1.5363 + 0.36 + 0.15 + 1.155 \\ Y &= 3.20 \text{ grams of fruit produced} \end{aligned}$$

If you want to learn more about the theory behind multiple regression, see Keller (2009) and Gould and Gould (2002).

7.4 Using Excel to Create a Correlation Matrix in Multiple Regression

The final step in multiple regression is to find the correlation between all of the variables that appear in the regression equation.

In our example, this means that we need to find the correlation between each of the six pairs of variables:

To do this, we need to use Excel to create a "correlation matrix." This matrix summarizes the correlations between all of the variables in the problem.

> Objective: To use Excel to create a correlation matrix between the four variables in this example.

To use Excel to do this, use these steps:
Data (top of screen under "Home" at the top left of screen)
Data Analysis

Correlation (scroll *up* to highlight this formula; see Fig. 7.6)

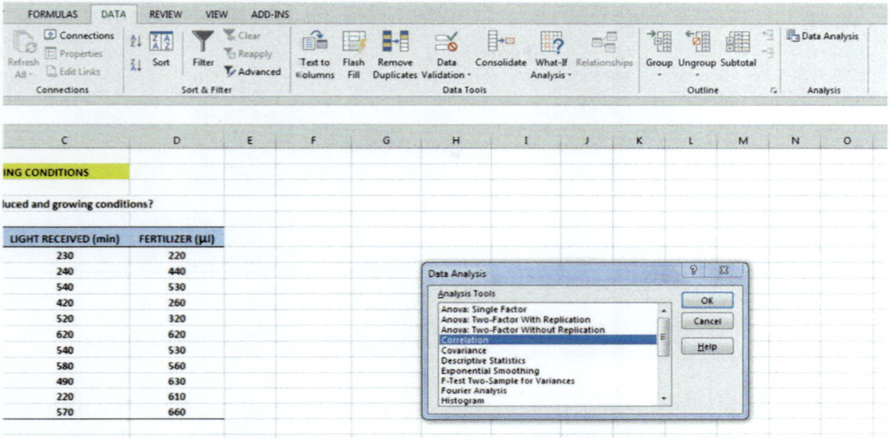

Fig. 7.6 Dialogue Box for Growing Conditions vs. Fruit Produced Correlations

OK

Input range: A6:D17

(Note that this input range includes the labels at the top of the FOUR variables (FRUIT PRODUCED, WATER, LIGHT RECEIVED, and FERTILIZER) as well as all of the figures in the original data set.)

Grouped by: Columns

Put a check in the box for: Labels in the First Row (since you included the labels at the top of the columns in your input range of data above)

Output range (click on the button to its left, and enter): A42 (see Fig. 7.7)

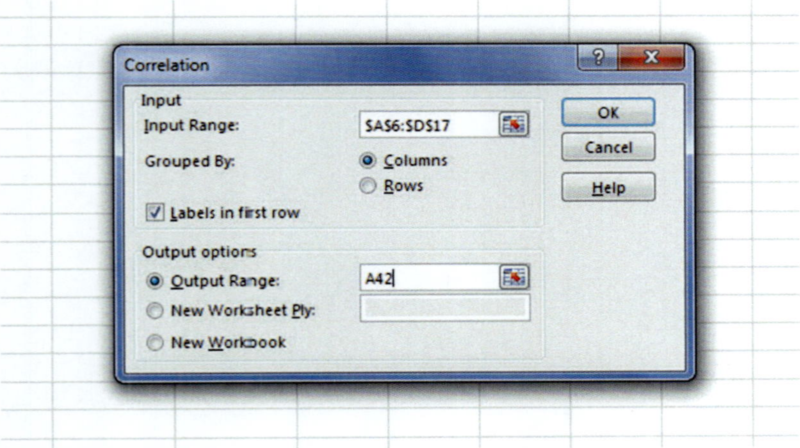

Fig. 7.7 Dialogue Box for Input/Output Range for Correlation Matrix

OK
The resulting correlation matrix appears in A42:E46 (See Fig. 7.8).

	A	B	C	D	E	
36	Intercept	1.5363	0.47	3.28	0.01	(
37	WATER (ml)	0.0006	0.00	0.67	0.53	(
38	LIGHT RECEIVED (min)	0.0003	0.00	0.30	0.78	(
39	FERTILIZER (µl)	0.0021	0.00	2.48	0.04	(
40						
41						
42		*FRUIT PRODUCED (g)*	*WATER (ml)*	*LIGHT RECEIVED (min)*	*FERTILIZER (µl)*	
43	FRUIT PRODUCED (g)	1				
44	WATER (ml)	0.510369686	1			
45	LIGHT RECEIVED (min)	0.446857676	0.468105152	1		
46	FERTILIZER (µl)	0.772523347	0.444074496	0.429202393	1	
47						

Fig. 7.8 Resulting Correlation Matrix for Growing Conditions vs. Fruit Produced Data

Next, format all of the numbers in the correlation matrix that are in decimals to two decimals places. And, also, make the columns wider so that all of the labels fit inside their cells. Then, center all the correlations in their cells.

Save this Excel file as: FRUIT5

The final spreadsheet for these scores appears in Fig. 7.9.

FRUIT PRODUCED IN RELATION TO GROWING CONDITIONS

Is there a relationship between fruit produced and growing conditions?

FRUIT PRODUCED (g)	WATER (ml)	LIGHT RECEIVED (min)	FERTILIZER (μl)
2.55	250	[illegible]30	220
3.05	610	[illegible]40	440
3.55	620	[illegible]40	530
2.05	420	[illegible]20	260
2.45	320	[illegible]20	320
2.95	630	[illegible]20	620
3.15	650	[illegible]40	530
3.45	520	[illegible]80	560
3.30	420	[illegible]90	630
2.75	330	[illegible]20	610
3.65	440	[illegible]70	660

SUMMARY OUTPUT

Regression Statistics	
Multiple R	0.80
R Square	0.64
Adjusted R Square	0.48
Standard Error	0.36
Observations	11

ANOVA

	df	*SS*	*MS*	*F*	*Significance F*
Regression	3	1.60	0.53	4.08	0.06
Residual	7	0.91	0.13		
Total	10	2.51			

	Coefficients	*Standard Error*	*t Stat*	*P-value*	*Lower 95%*	*Upper 95%*	*Lower 95.0%*	*Upper 95.0%*
Intercept	1.5363	0.47	3.28	0.01	0.43	2.64	0.43	2.64
WATER (ml)	0.0006	0.00	0.67	0.53	0.00	0.00	0.00	0.00
LIGHT RECEIVED (min)	0.0003	0.00	0.30	0.78	0.00	0.00	0.00	0.00
FERTILIZER (μl)	0.0021	0.00	2.48	0.04	0.00	0.00	0.00	0.00

FRUIT4

	FRUIT PRODUCED (g)	*WATER (ml)*	*LIGHT RECEIVED (min)*	*FERTILIZER (μl)*
FRUIT PRODUCED (g)	1			
WATER (ml)	0.51	1		
LIGHT RECEIVED (min)	0.45	0.47	1	
FERTILIZER (μl)	0.77	0.44	0.43	1

Fig. 7.9 Final Spreadsheet for Growing Conditions vs. Fruit Produced Regression and the Correlation Matrix

Note that the number "1" along the diagonal of the correlation matrix means that the correlation of each variable with itself is a perfect, positive correlation of 1.0.

Correlation coefficients are always expressed in just two decimal places.

You are now ready to read the correlation between the six pairs of variables:

The correlation between WATER and FRUIT PRODUCED is :	*+.51*
The correlation between LIGHT RECEIVED and FRUIT PRODUCED is :	*+.45*
The correlation between FERTILIZER and FRUIT PRODUCED is :	*+.77*
The correlation between LIGHT RECEIVED and WATER is :	*+.47*
The correlation between FERTILIZER and WATER is :	*+.44*
The correlation between FERTILIZER and LIGHT RECEIVED is :	*+.43*

This means that the best predictor of FRUIT PRODUCED is FERTILIZER with a correlation of +.77. Adding the other two predictor variables, WATER and LIGHT RECEIVED, improved the prediction by only 0.03 to 0.80, and was, therefore, only slightly better in prediction. FERTILIZER is an excellent predictor of FRUIT PRODUCED by itself.

If you want to learn more about the correlation matrix, see Levine et al. (2011).

7.5 End-of-Chapter Practice Problems

1. Agriculture around the world depends on viable seed production. Crops that originated in various parts of the world are now grown in a wide range of climates. For example, corn (maize) originated in present day Mexico and is now grown throughout the world. The amount of seeds produced in different growing conditions, especially in marginal climates, affect how successful a plant will be for agricultural purposes. The main conditions that can affect plant growth are water, light, and fertilizer. Additionally, "plant crowding" in terms of how close the plants are to each other generally causes negative changes in productivity (i.e., the closer the plants are to one another, the fewer the seeds produced). As a result, agricultural companies have started studying the impact that space between plants has on overall plant production.

   Suppose that you have been asked by a large agricultural seed supplier to analyze some data from their experimental greenhouse. They have asked you to look at how growing conditions predict the amount of seed production for a new type of agricultural plant. You have been asked to use the following growing variables: water (milliliters, ml), light received (in minutes, min), and fertilizer (microliters, μl), and, in addition, to use space (centimeters, cm) between plants as an additional predictor of seed production. The seed supplier has asked you for your recommendation as to whether or not space should be included with the main growing conditions.

   You have decided to use a multiple correlation and multiple regression analysis, and to test your Excel skills, you have collected the data from a random sample of seed pods from 12 crop plants that have been grown under controlled conditions during one full growing season. These hypothetical data appear in Fig. 7.10:

SEED PRODUCTION				
How well do water, light, fertilizer, and space predict the number of seeds produced per pod?				
AVERAGE SEEDS PER POD	WATER (ml)	LIGHT RECEIVED (min)	SPACE (cm)	FERTILIZER (µl)
3.25	600	620	12.5	650
3.42	520	550	10	600
2.85	510	540	5	500
2.65	480	460	2.5	510
3.65	720	710	15	630
3.16	570	610	7.5	550
3.56	710	650	10	610
2.35	500	480	5	430
2.86	450	470	7.5	450
2.95	560	530	10	550
3.15	550	580	10	580
3.45	610	620	12.5	620

Fig. 7.10 Worksheet Data for Chap. 7: Practice Problem #1

(a) Create an Excel spreadsheet using AVERAGE SEEDS PER POD as the criterion (Y), and WATER (X_1), LIGHT RECEIVED (X_2), SPACE (X_3), and FERTILIZER (X_4) as the predictors.
(b) Use Excel's *multiple regression* function to find the relationship between these five variables and place it below the table.
(c) Use number format (two decimal places for the multiple correlation on the SUMMARY OUTPUT, and use four decimal places for the coefficients in the SUMMARY OUTPUT).
(d) Print the table and regression results below the table so that they fit onto one page.
(e) Save this file as: seed14

Answer the following questions using your Excel printout:

1. What is the multiple correlation R_{xy}?
2. What is the y-intercept *a*?
3. What is the coefficient for WATER b_1?
4. What is the coefficient for LIGHT RECEIVED b_2?
5. What is the coefficient for SPACE b_3?
6. What is the coefficient for FERTILIZER b_4?
7. What is the multiple regression equation?
8. Predict the AVERAGE SEEDS PER POD you would expect for a WATER score of 610, a LIGHT RECEIVED score of 550, a SPACE score of 3, and a FERTILIZER score of 610.

(f) Now, go back to your Excel file and create a *correlation matrix* for these five variables, and place it underneath the SUMMARY OUTPUT.
(g) Save this file as: seed15
(h) Now, print out *just this correlation matrix* on a separate sheet of paper.

Answer the following questions using your Excel printout. Be sure to include the plus or minus sign for each correlation:

9. What is the correlation between WATER and AVERAGE SEEDS PER POD?
10. What is the correlation between LIGHT RECEIVED and AVERAGE SEEDS PER POD?
11. What is the correlation between SPACE and AVERAGE SEEDS PER POD?
12. What is the correlation between FERTILIZER and AVERAGE SEEDS PER POD?
13. What is the correlation between SPACE and WATER?
14. What is the correlation between LIGHT RECEIVED and FERTILIZER?
15. Discuss which of the four predictors is the best predictor of AVERAGE SEEDS PER POD.
16. Explain in words how much better the four predictor variables together predict AVERAGE SEEDS PER POD than the best single predictor by itself.

2. In order to grow properly, most plants need water, warmth, carbon dioxide gas, light, and minerals. Suppose you wanted to study the relationship between temperature and precipitation and their effect on plant productivity. In ecological terms, "productivity" refers to the amount of plant growth and it is measured in grams per meter squared per year (g/m^2/year) at the site. Precipitation (rainfall) is measured as the annual mean precipitation (cm/year) at the site. Temperature is measured as the average annual temperature in degrees Centigrade (°C) at the site. Let productivity be the dependent variable (criterion), and precipitation and temperature be the independent variables (predictors) at each site. Hypothetical data for 14 sites are presented in Fig. 7.11.

	PLANT-GROWTH DATA	
PRODUCTIVITY (g/m squared/year)	MEAN ANNUAL PRECIPITATION (cm/year)	MEAN ANNUAL TEMPERATURE (degrees C)
250	100	-13
300	200	2.5
400	100	1.5
350	200	2.5
450	300	0
550	400	-2.5
500	200	-4
650	200	-6.5
600	400	-2
750	200	4.5
800	400	-9.5
850	500	2
950	200	3
1000	500	3

Fig. 7.11 Worksheet Data for Chap. 7: Practice Problem #2

(a) create an Excel spreadsheet using PRODUCTIVITY as the criterion (Y), and the other variables as the two predictors of this criterion.
(b) Use Excel's *multiple regression* function to find the relationship between these variables and place it below the table.
(c) Use number format (two decimal places) for the multiple correlation on the Summary Output, and use number format (three decimal places) for the coefficients in the Summary Output.
(d) Print the table and regression results below the table so that they fit onto one page.
(e) By hand on this printout, *circle and label:*

(1a) multiple correlation R_{xy}
(2b) coefficients for the y-intercept, precipitation, and temperature.

(f) Save this file as: PLANT3A
(g) Now, go back to your Excel file and create a correlation matrix for these three variables, and place it underneath the Summary Table. *Change each correlation to just two decimals.* Save this file again as: PLANT3A
(h) Now, print out *just this correlation matrix in portrait mode* on a separate sheet of paper.

Answer the following questions using your Excel printout:

1. What is the multiple correlation R_{xy}?
2. What is the y-intercept *a*?
3. What is the coefficient for precipitation b_1?
4. What is the coefficient for temperature b_2?
5. What is the multiple regression equation?
6. Underneath this regression equation by hand, predict the productivity you would expect for an annual precipitation of 300 cm/yr and an annual temperature of +2 degrees Centigrade.

Answer the following questions using your Excel printout. Be sure to include the plus or minus sign for each correlation:

7. What is the correlation between precipitation and productivity?
8. What is the correlation between temperature and productivity?
9. What is the correlation between temperature and precipitation?
10. Discuss which of the two predictors is the better predictor of productivity.
11. Explain in words how much better the two predictor variables combined predict productivity than the better single predictor by itself.

3. The Advanced Placement (AP) Tests are standardized tests that allow high school students to "test out" of college courses, either by receiving credit for these courses or by allowing the students to waive courses in which their AP scores are very high. The AP Exams are scored on a 5-point scale in which a score of "1" means "No recommendation" and a score of "5" means "Extremely well qualified." There are 34 AP courses for which these exams are available. The Biology Test is intended to measure the equivalent of a college introductory

biology course taken by Biology Majors during their first year of college; the test has three general areas: (1) Molecules and Cells, (2) Heredity and Evolution, and (3) Organisms and Populations.

Suppose that you have been asked by the Chair of the Biology Department at a selective university to see how well the AP Biology Test, along with other predictors, predicts GPA at the end of the first year of study for Biology majors. This Chair has asked you to use as predictors: (1) SAT-Verbal scores, (2) - SAT-Math scores, and (3) high school GPA, and, in addition, to use the AP Biology Test score as an additional predictor of this GPA. The Chair would like your recommendation as to whether or not the AP Biology Test should become an admissions requirement in addition to the SAT for admission to the undergraduate major in Biology.

You have decided to use a multiple correlation and multiple regression analysis, and to test your Excel skills, you have collected the data of a random sample of ten Biology students who have just finished their first year of study at this university. These hypothetical data appear in Fig. 7.12.

FROSH GPA	SAT-VERBAL	SAT-MATH	HS GPA	AP BIOLOGY TEST
3.23	650	510	3.55	2
2.90	490	420	2.96	2
2.80	630	540	3.27	5
3.42	520	520	3.45	4
2.80	560	540	2.90	3
2.90	410	420	2.80	2
2.35	450	460	2.63	2
2.58	420	430	2.71	1
3.12	560	520	3.26	3
3.47	650	680	3.45	4

Fig. 7.12 Worksheet Data for Chap. 7: Practice Problem #3

(a) create an Excel spreadsheet using FROSH GPA as the criterion and the other four variables as the predictors.
(b) Use Excel's *multiple regression* function to find the relationship between these five variables and place the SUMMARY OUTPUT below the table.
(c) Use number format (two decimal places) for the multiple correlation on the Summary Output, and use number format (three decimal places) for the coefficients in the summary output and for all other decimal figures in the SUMMARY OUTPUT.

(d) Save the file as:
APbiology
(e) Print the table and regression results below the table so that they fit onto one page.

Answer the following questions using your Excel printout:

1. What is multiple correlation R_{xy}?
2. What is the y-intercept a?
3. What is the coefficient for SAT-V b_1?
4. What is the coefficient for SAT-M b_2?
5. What is the coefficient for HS GPA b_3**?**
6. What is the coefficient for the AP BIOLOGY b_4?
7. What is the multiple regression equation?
8. Predict the FROSH GPA you would expect for an SAT-V score of 650, an SAT-M score of 630, a HS GPA of 3.47, and an AP Biology Test score of 4.

(f) Now, go back to your Excel file and create a correlation matrix for these five variables, and place it underneath the SUMMARY OUTPUT on your spreadsheet.
(g) Save this file as: APbiology3
(h) Now, print out *just this correlation matrix* on a separate sheet of paper.

Answer the following questions using your Excel printout. Be sure to include the plus or minus sign for each correlation:

9. What is the correlation between SAT-V and FROSH GPA?
10. What is the correlation between SAT-M and FROSH GPA?
11. What is the correlation between HS GPA and FROSH GPA?
12. What is the correlation between AP BIOLOGY and FROSH GPA?
13. What is the correlation between SAT-V and SAT-M?
14. What is the correlation between HS GPA and AP-BIOLOGY?
15. What is the correlation between SAT-M and AP BIOLOGY?
16. What is the correlation between SAT-V and AP BIOLOGY?
17. Discuss which of the four predictors is the best predictor of FROSH GPA.
18. Explain in words how much better the four predictor variables combined predict FROSH GPA than the best single predictor by itself.

References

Gould, J.L. and Gould, G.F. Biostats Basics: A Student Handbook. New York, NY: W.H. Freeman and Company, 2002.

Keller, G. Statistics for Management and Economics (8th ed.). Mason, OH: South-Western Cengage Learning, 2009.

Levine, D.M., Stephan, D.F., Krehbiel, T.C., and Berenson, M.L. Statistics for Managers using Microsoft Excel (6th ed.). Boston, MA: Prentice Hall/Pearson, 2011.

Chapter 8
One-Way Analysis of Variance (ANOVA)

So far in this 2016 Excel Guide, you have learned how to use a one-group t-test to compare the sample mean to the population mean, and a two-group t-test to test for the difference between two sample means. *But what should you do when you have more than two groups and you want to determine if there is a significant difference between the means of these groups?*

The answer to this question is: *Analysis of Variance (ANOVA).*

The ANOVA test allows you to test for the difference between the means when you have *three or more groups* in your research study.

Important note: In order to do One-way Analysis of Variance, you need to have installed the "Data Analysis Toolpak" that was described in Chap. 6 (see Sect. 6.5.1). If you did not install this, you need to do that now.

Let's suppose that you wanted to study the sound bursts of three subspecies of forager honey bees in relation to a food source that was 400 meters from their hive. These honey bees "dance" by producing sound bursts in a cycle, and the researchers have obtained the average number of sound bursts per cycle for these bees. The hypothetical results are given in Fig. 8.1. Is there any evidence that these three subspecies of honey bees produce a different number of sound bursts per cycle?

T.J. Quirk et al., *Excel 2016 for Biological and Life Sciences Statistics*, Excel for Statistics, DOI 10.1007/978-3-319-39489-3_8

Fig. 8.1 Worksheet Data for Honey Bees (Practical Example)

SOUND BURSTS OF DANCING FORAGER HONEY BEES		
SUBSPECIES A	SUBSPECIES B	SUBSPECIES C
14.2	13.2	15
16.5	15.3	17
15.7	16.6	18
17.1	14.1	16
16.6	16.6	18
18.3	15.2	15
14.2	13.3	17
16.7	14.1	16
15.8	13.3	18
18.5	14.9	16
17.8		17
		18

You have been asked to analyze the data to determine if there was any significant difference in the average number of sound bursts per cycle between the three subspecies of honey bees. To test your Excel skills, you have selected a random sample of bees from each of these subspecies (see Fig. 8.1). Note that each subspecies can have a different number of bees in order for ANOVA to be used on the data. Statisticians delight in this fact by referring to this characteristic by stating that: "ANOVA is a very robust test." (Statisticians love that term!)

Create an Excel spreadsheet for these data in this way:

A3: SOUND BURSTS OF DANCING FORAGER HONEY BEES
A5: SUBSPECIES A
B5: SUBSPECIES B
C5: SUBSPECIES C
A6: 14.2
C17: 18

Enter the other information into your spreadsheet table. When you have finished entering these data, the last cell on the left should have 17.8 in cell A16, and the last cell on the right should have 18 in cell C17. Center the numbers in each of the columns.

Important note: Be sure to double-check all of your figures in the table to make sure that they are exactly correct or you will not be able to obtain the correct answer for this problem!

Save this file as: BEE10

8.1 Using Excel to Perform a One-Way Analysis of Variance (ANOVA)

Objective: To use Excel to perform a one-way ANOVA test.

You are now ready to perform an ANOVA test on these data using the following steps:

Data (at top of screen)
Data Analysis (far right at top of screen)
Anova: Single Factor (*scroll up to this formula and highlight it*; see Fig. 8.2)

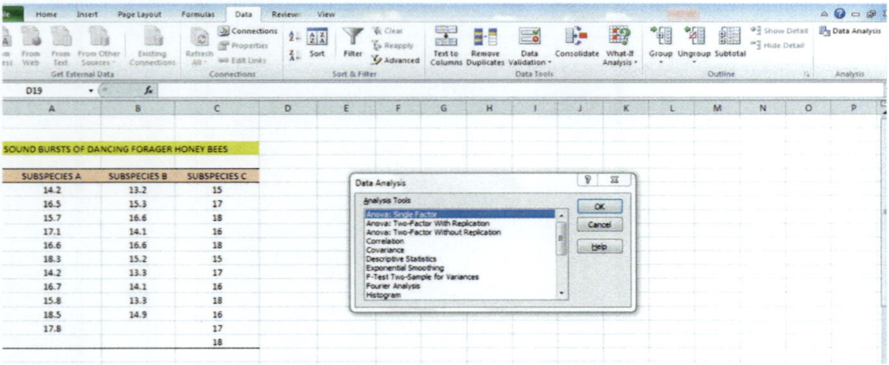

Fig. 8.2 Dialog Box for Data Analysis: Anova Single Factor

OK
Input range: A5: C17 (note that you have included in this range the column titles that are in row 5)

Important note: *Whenever the data set has a different sample size in the groups being compared, the INPUT RANGE that you define must start at the column title of the first group on the left and go to the last column on the right to the lowest row that has a figure in it in the entire data matrix so that the INPUT RANGE has the "shape" of a rectangle when you highlight it. Since SUBSPECIES C has 18 in cell C17, your "rectangle" must include row 17!*

Grouped by: Columns
Put a check mark in: Labels in First Row
Output range (click on the button to its left): A20 (see Fig. 8.3)

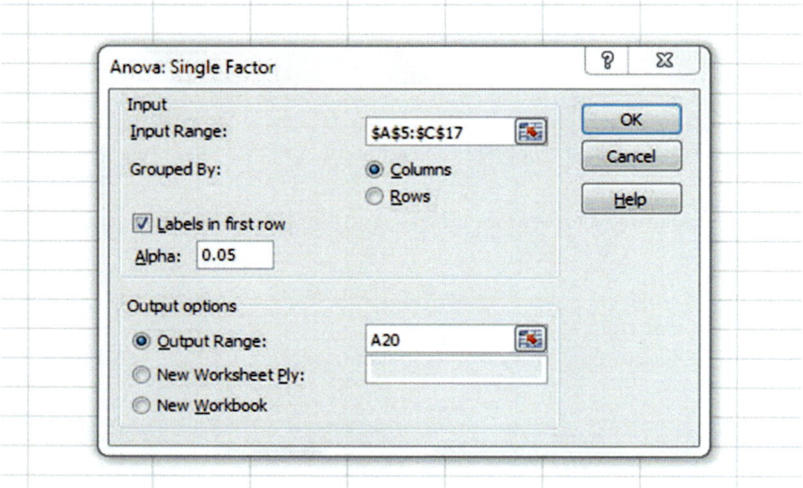

Fig. 8.3 Dialog Box for Anova: Single Factor Input/Output Range

OK

Save this file as: BEE11

You should have generated the table given in Fig. 8.4.

15	18.5	14.9	16				
16	17.8		17				
17			18				
18							
19							
20	Anova: Single Factor						
21							
22	SUMMARY						
23	*Groups*	*Count*	*Sum*	*Average*	*Variance*		
24	SUBSPECIES A	11	181.40	16.49	2.10		
25	SUBSPECIES B	10	146.60	14.66	1.64		
26	SUBSPECIES C	12	201.00	16.75	1.30		
27							
28							
29	ANOVA						
30	*Source of Variation*	*SS*	*df*	*MS*	*F*	*P-value*	*F crit*
31	Between Groups	27.33	2	13.66	8.19	0.001	3.32
32	Within Groups	50.04	30	1.67			
33							
34	Total	77.37	32				
35							

Fig. 8.4 ANOVA Results for Honey Bees

Center all of the numbers in the ANOVA table, and round off all numbers that are decimals to two decimal places.

Save this file as: BEE12

Print out both the data table and the ANOVA summary table so that all of this information fits onto one page. (Hint: Set the Page Layout/Fit to Scale to *85 % size*).

As a check on your analysis, you should have the following in these cells:

A20: Anova: Single Factor
D25: 14.66
B26: 12
D32: 1.67
E31: 8.19
G31: 3.32

Now, let's discuss how you should interpret this table:

8.2 How to Interpret the ANOVA Table Correctly

Objective: To interpret the ANOVA table correctly

ANOVA allows you to test for the differences between means when you have three or more groups of data. This ANOVA test is called the F-test statistic, and is typically identified with the letter: F.

The formula for the F-test is this:

F = Mean Square between groups (MS_b) divided by Mean Square within groups (MS_w)

$$F = MS_b/MS_w \tag{8.1}$$

The derivation and explanation of this formula is beyond the scope of this *Excel Guide*. In this *Excel Guide*, we are attempting to teach you *how to use Excel*, and we are not attempting to teach you the statistical theory that is behind the ANOVA formulas. For a detailed explanation of ANOVA, see Gould and Gould (2002) and Weiers (2011).

Note that cell D31 contains $MS_b = 13.66$, while cell D32 contains $MS_w = 1.67$.

When you divide these two figures using their cell references in Excel, you get the answer for the F-test of 8.19 which is in cell E31. (Remember, Excel is more accurate than your calculator!) Let's discuss now the meaning of the figure: F = 8.19.

In order to determine whether this figure for F of 8.19 indicates a significant difference between the means of the three groups, the first step is to write the null hypothesis and the research hypothesis for the three subspecies of honey bees.

In our statistics comparisons, the null hypothesis states that the population means of the three groups are equal, while the research hypothesis states that the population means of the three groups are not equal and that there is, therefore, a significant difference between the population means of the three groups. Which of these two hypotheses should you accept based on the ANOVA results?

8.3 Using the Decision Rule for the ANOVA F-Test

To state the hypotheses, let's call SUBSPECIES A as Group 1, SUBSPECIES B as Group 2, and SUBSPECIES C as Group 3. The hypotheses would then be:

H_0: $\mu_1 = \mu_2 = \mu_3$
H_1: $\mu_1 \neq \mu_2 \neq \mu_3$

The answer to this question is analogous to the decision rule used in this book for both the one-group t-test and the two-group t-test. You will recall that this rule (See Sects. 4.1.6 and 5.1.8) was:

If the absolute value of t is less than the critical t, you accept the null hypothesis.

or

If the absolute value of t is greater than the critical t, you reject the null hypothesis, and accept the research hypothesis.

Now, here is the decision rule for ANOVA:

Objective: To learn the decision rule for the ANOVA F-test

The decision rule for the ANOVA F-test is the following:

If the value for F is less than the critical F-value, accept the null hypothesis.

or

If the value of F is greater than the critical F-value, reject the null hypothesis, and accept the research hypothesis.

Note that Excel tells you the critical F-value in cell G31: 3.32
Therefore, our decision rule for the AVOVA test is this:

Since the value of F of 8.19 is greater than the critical F-value of 3.32, we reject the null hypothesis and accept the research hypothesis.

Therefore, our conclusion, in plain English, is:

There is a significant difference between the average number of sound bursts per cycle between the three subspecies of honey bees.

Note that it is not necessary to take the absolute value of F of 8.19. The F-value can never be less than one, and so it can never be a negative value which requires us to take its absolute value in order to treat it as a positive value.

It is important to note that ANOVA tells us that there was a significant difference between the population means of the three groups, *but it does not tell us which pairs of groups were significantly different from each other.*

8.4 Testing the Difference Between Two Groups Using the ANOVA t-Test

To answer that question, we need to do a different test called the ANOVA t-test.

Objective:	To test the difference between the means of two groups using an ANOVA t-test when the ANOVA results indicate a significant difference between the population means.

Since we have three groups of data (one group for each of the three subspecies of bees), we would have to perform three separate ANOVA t-tests to determine which pairs of groups were significantly different. This requires that we would have to perform a separate ANOVA t-test for the following pairs of groups:

1. SUBSPECIES A vs. SUBSPECIES B
2. SUBSPECIES A vs. SUBSPECIES C
3. SUBSPECIES B vs. SUBSPECIES C

We will do just one of these pairs of tests, SUBSPECIES B vs. SUBSPECIES C, to illustrate the way to perform an ANOVA t-test comparing these two subspecies. The ANOVA t-test for the other two pairs of groups would be done in the same way.

8.4.1 Comparing Subspecies B vs. Subspecies C in the Number of Sound Bursts per Cycle Using the ANOVA t-Test

Objective:	To compare Subspecies B vs. Subspecies C in number of sound bursts per cycle using the ANOVA t-test.

The first step is to write the null hypothesis and the research hypothesis for these two subspecies of bees.

For the ANOVA t-test, the null hypothesis is that the population means of the two groups are equal, while the research hypothesis is that the population means of the two groups are not equal (i.e., there is a significant difference between these two means). Since we are comparing SUBSPECIES B (Group 2) vs. SUBSPECIES C (Group 3), these hypotheses would be:

H_0: $\mu_2 = \mu_3$
H_1: $\mu_2 \neq \mu_3$

For Group 2 vs. Group 3, the formula for the ANOVA t-test is:

$$ANOVA\ \ t = \frac{\overline{X}_1 - \overline{X}_2}{s.e._{ANOVA}} \tag{8.2}$$

where

$$s.e._{ANOVA} = \sqrt{MS_w\left(\frac{1}{n_1} + \frac{1}{n_2}\right)} \tag{8.3}$$

The steps involved in computing this ANOVA t-test are:

1. Find the difference of the sample means for the two groups ($14.66 - 16.75 = -2.09$).
2. Find $1/n_2 + 1/n_3$ (since both groups have a different number of bees in them, this becomes: $1/10 + 1/12 = 0.10 + 0.08 = 0.18$)
3. Multiply MS_w times the answer for step 2 ($1.67 \times 0.18 = 0.31$)
4. Take the square root of step 3 (SQRT $(0.31) = 0.55$)
5. Divide Step 1 by Step 4 to find ANOVA t ($-2.09/0.55 = -3.80$)

Note: Since Excel computes all calculations to 16 decimal places, when you use Excel for the above computations, your answer will be −3.78 in two decimal places, but Excel's answer will be much more accurate because it is always in 16 decimal places.

Now, what do we do with this ANOVA t-test result of −3.80? In order to interpret this value of −3.80 correctly, we need to determine the critical value of t for the ANOVA t-test. To do that, we need to find the degrees of freedom for the ANOVA t-test as follows:

8.4.1.1 Finding the Degrees of Freedom for the ANOVA t-test

Objective: To find the degrees of freedom for the ANOVA t-test.

The degrees of freedom (df) for the ANOVA t-test is found as follows:

df = take the total sample size of all of the groups and subtract the number of groups in your study ($n_{TOTAL} - k$ where k = the number of groups)

In our example, the total sample size of the three groups is 33 since there are 11 bees in Group 1, 10 bees in Group 2, and 12 bees in Group 3, and since there are three groups, 33 − 3 gives a degrees of freedom for the ANOVA t-test of 30.

If you look up df = 30 in the t-table in Appendix E in the degrees of freedom column (df), which is the *second column on the left of this table*, you will find that the critical t-value is 2.042.

Important note: Be sure to use the degrees of freedom column (df) in Appendix E for the ANOVA t-test critical t value

8.4.1.2 Stating the Decision Rule for the ANOVA t-test

Objective: To learn the decision rule for the ANOVA t-test

Interpreting the result of the ANOVA t-test follows the same decision rule that we used for both the one-group t-test (see Sect. 4.1.6) and the two-group t-test (see Sect. 5.1.8):

If the absolute value of t is less than the critical value of t, we accept the null hypothesis.

or

If the absolute value of t is greater than the critical value of t, we reject the null hypothesis and accept the research hypothesis.

Since we are using a type of t-test, we need to take the absolute value of t. Since the absolute value of −3.80 is greater than the critical t-value of 2.042, we reject the null hypothesis (that the population means of the two groups are equal) and accept the research hypothesis (that the population means of the two groups are significantly different from one another).

This means that our conclusion, in plain English, is as follows:

The average number of sound bursts per cycle for SUBSPECIES C was significantly greater than the average number of sound bursts per cycle for SUBSPECIES B (16.75 vs. 14.66).

8.4.1.3 Performing an ANOVA t-Test Using Excel Commands

Now, let's do these calculations for the ANOVA t-test using Excel with the file you created earlier in this chapter: BEE12

A36 : SUBSPECIES B vs. SUBSPECIES C
A38 : 1/n B + 1/n C
A40 : s.e. B vs. C
A42 : ANOVA t-test
B38 : = (1/10 + 1/12) (no spaces between symbols)
B40 : = SQRT(D32*B38)
B42 : = (D25 – D26)/B40 (no spaces between symbols)

You should now have the following results in these cells when you round off all these figures in the ANOVA t-test to two decimal points.:

B38 : 0.18
B40 : 0.55
B42 : – 3.78

Save this final result under the file name: BEES6

Print out the resulting spreadsheet so that it fits onto one page like Fig. 8.5 (Hint: Reduce the Page Layout/Scale to Fit to *85 %*).

SOUND BURSTS OF DANCING FORAGER HONEY BEES

SUBSPECIES A	SUBSPECIES B	SUBSPECIES C
14.2	13.2	15
16.5	15.3	17
15.7	16.6	18
17.1	14.1	16
16.6	16.6	18
18.3	15.2	15
14.2	13.3	17
16.7	14.1	16
15.8	13.3	18
18.5	14.9	16
17.8		17
		18

Anova: Single Factor

SUMMARY

Groups	*Count*	*Sum*	*Average*	*Variance*
SUBSPECIES A	11	181.40	16.49	2.10
SUBSPECIES B	10	146.60	14.66	1.64
SUBSPECIES C	12	201.00	16.75	1.30

ANOVA

Source of Variation	*SS*	*df*	*MS*	*F*	*P-value*	*F crit*
Between Groups	27.33	2	13.66	8.19	0.001	3.32
Within Groups	50.04	30	1.67			
Total	77.37	32				

SUBSPECIES B vs. SUBSPECIES C	
1/n B + 1/n C	0.18
s.e. B vs. C	0.55
ANOVA t-test	-3.78

Fig. 8.5 Final Spreadsheet of Honey Bees

For a more detailed explanation of the ANOVA t-test, see Black (2010).

Important note: You are only allowed to perform an ANOVA t-test comparing the population means of two groups when the F-test produces a significant difference between the population means of all of the groups in your study.

*It is improper to do any ANOVA t-test when the value of F is less than the critical value of F. Whenever F is less than the critical F, this means that there was no difference between the population means of the groups, and, therefore, that you cannot test to see if there is a difference between the means of any two groups since this would capitalize on chance differences between these two groups. For more information on this important point, see Gould and Gould (*2002*).*

8.5 End-of-Chapter Practice Problems

1. In a controlled greenhouse experiment, what type of soil conditions are favorable for the growth of a common flowering species such as Daisy Fleabane (*Erigeron annuus*) often found along roadsides? Suppose that you wanted to conduct a controlled research study to compare the growth of this flower in three types of soil treatments, specifically the effects of Nitrogen. Nitrogen is an essential macro-nutrient necessary for plant development and is commonly found in fertilizers: (1) Group 1: No nitrogen added to the soil, (2) Group 2: Low amounts of Nitrogen added to the soil, and (3) Group 3: High amounts of Nitrogen added to the soil. Your measured variable will be milligrams (mg) of dry biomass of *Erigeron annuus*. To do this study, you have obtained soil from one location of land in the state of Colorado and seeds from a distributor. You have sterilized the soil so that you could place it in 12 containers for each type of treatment (i.e., 36 containers in all). Suppose, further, that you planted the same number of seeds of *Erigeron annuus* in each container and maintained the containers in a controlled greenhouse under constant conditions for a growing season. You then clipped and dried all the *Erigeron annuus* from each treatment to obtain dry biomass. You then weighed the dry biomass of each treatment and generated the hypothetical data given in Fig. 8.6. Note that there are a different number of containers in each of the three treatments because some of the plants did not survive the growing season. To test your Excel skills, you have decided to run an ANOVA test of the data to determine the growing characteristics of this plant with respect to varying amounts of Nitrogen.

DAISY FLEABANE GROWTH IN ELEVATED NITROGEN SOILS		
No Nitrogen	Low Nitrogen	High Nitrogen
500	550	800
550	600	1000
700	750	900
650	700	1100
500	600	1400
550	650	1600
600	600	1500
650	700	1800
600	650	1500
	750	1800
		2000
		1600

Fig. 8.6 Worksheet Data for Chap. 8: Practice Problem #1

(a) Enter these data on an Excel spreadsheet.
(b) Perform a *one-way ANOVA test* on these data, and show the resulting ANOVA table *underneath* the input data for the three levels of Nitrogen.
(c) If the F-value in the ANOVA table is significant, create an Excel formula to compute the ANOVA t-test comparing the average for Low Nitrogen against High Nitrogen and show the results below the ANOVA table on the spreadsheet (put the standard error and the ANOVA t-test value on separate lines of your spreadsheet, and use two decimal places for each value)
(d) Print out the resulting spreadsheet so that all of the information fits onto one page
(e) Save the spreadsheet as: Weed5

Now, write the answers to the following questions using your Excel printout:

1. What are the null hypothesis and the research hypothesis for the ANOVA F-test?
2. What is MS_b on your Excel printout?
3. What is MS_w on your Excel printout?
4. Compute $F = MS_b/MS_w$ using your calculator.
5. What is the critical value of F on your Excel printout?
6. What is the result of the ANOVA F-test?
7. What is the conclusion of the ANOVA F-test in plain English?
8. If the ANOVA F-test produced a significant difference between the three amounts of Nitrogen in the weight of the flowers grown in the greenhouse during a growing season, what is the null hypothesis and the research hypothesis for the ANOVA t-test comparing Low Nitrogen versus High Nitrogen?

9. What is the mean (average) for Low Nitrogen on your Excel printout?
10. What is the mean (average) for High Nitrogen on your Excel printout?
11. What are the degrees of freedom (df) for the ANOVA t-test comparing Low Nitrogen versus High Nitrogen?
12. What is the critical t value for this ANOVA t-test in Appendix E for these degrees of freedom?
13. Compute the $s.e._{ANOVA}$ using your calculator.
14. Compute the ANOVA t-test value comparing Low Nitrogen versus High Nitrogen using your calculator.
15. What is the result of the ANOVA t-test comparing Low Nitrogen versus High Nitrogen?
16. What is the conclusion of the ANOVA t-test comparing Low Nitrogen versus High Nitrogen in plain English?

Note that since there are three treatments, you need to do three ANOVA t-tests to determine what the significant differences are between the three treatments. *Since you have just completed the ANOVA t-test comparing Low Nitrogen versus High Nitrogen, you would also need to do the ANOVA t-test comparing Low Nitrogen versus No Nitrogen, and also the ANOVA t-test comparing No Nitrogen versus High Nitrogen in order to write a conclusion summarizing all three of these three types of ANOVA t-tests.*

2. College students take Introductory Biology for a variety of reasons. Some students take it because it is a required Premed course, others take it because they are Biology majors, some take it because they are Science majors other than Biology, and some students take it as a General Education all-college required course. Suppose that you wanted to compare the cumulative GPAs of students who are taking an Introductory Biology course this semester in terms of their reasons for taking this course. Also, you have decided to exclude students who scored a 5 on the Advancement Placement Science Exam from your research study since these students receive academic credit for this course without having to take this course. You decide to take a random sample of students taking Introductory Biology this semester to test your Excel skills, and the hypothetical data are given in Fig. 8.7.

INTRODUCTORY BIOLOGY: REASONS FOR TAKING THIS COURSE			
	Cumulative GPAs when entering this course		
A	B	C	D
Premed required course	Biology major	Other Science major	General Ed requirement
2.84	2.60	2.50	2.00
3.26	2.86	2.65	2.10
3.05	2.95	2.70	2.25
3.15	3.00	2.55	2.35
3.35	3.14	2.80	2.43
3.40	3.25	3.20	2.67
3.55	3.35	3.35	3.00
3.68	3.60	3.50	3.16
3.84	3.70	3.60	3.26
3.90		3.70	3.55
		3.80	3.35
			3.25

Fig. 8.7 Worksheet Data for Chap. 8: Practice Problem #2

(a) Enter these data on an Excel spreadsheet.
(b) Perform a *one-way ANOVA test* on these data, and show the resulting ANOVA table *underneath* the input data for the four types of students. Round off all decimal figures to two decimal places, and center all numbers in the ANOVA table.
(c) If the F-value in the ANOVA table is significant, create an Excel formula to compute the ANOVA t-test comparing the GPA for Premed students against the GPA for Other Science majors, and show the results below the ANOVA table on the spreadsheet (put the standard error and the ANOVA t-test value on separate lines of your spreadsheet, and use two decimal places for each value)
(d) Print out the resulting spreadsheet so that all of the information fits onto one page
(e) Save the spreadsheet as: BIO3

Let's call the Premed students Group A, the Biology majors Group B, the Other Science majors Group C and the General Ed requirement students Group D. *Now, write the answers to the following questions using your Excel printout:*

1. What are the null hypothesis and the research hypothesis for the ANOVA F-test?
2. What is MS_b on your Excel printout?
3. What is MS_w on your Excel printout?
4. Compute $F = MS_b/MS_w$ using your calculator.
5. What is the critical value of F on your Excel printout?
6. What is the result of the ANOVA F-test?
7. What is the conclusion of the ANOVA F-test in plain English?

8. If the ANOVA F-test produced a significant difference between the four types of students in their cumulative GPAs when entering the Introductory Biology course, what is the null hypothesis and the research hypothesis for the ANOVA t-test comparing Premed students (Group A) versus Other Science majors (Group C)?
9. What is the mean (average) GPA for Premed students on your Excel printout?
10. What is the mean (average) GPA for Other Science majors on your Excel printout?
11. What are the degrees of freedom (df) for the ANOVA t-test comparing Premed students versus Other Science majors?
12. What is the critical t value for this ANOVA t-test in Appendix E for these degrees of freedom?
13. Compute the $s.e._{ANOVA}$ using Excel for Premed students versus Other Science majors.
14. Compute the ANOVA t-test value comparing Premed students versus Other Science majors using Excel.
15. What is the result of the ANOVA t-test comparing Premed students versus Other Science majors?
16. What is the conclusion of the ANOVA t-test comparing Premed students versus Other Science majors in plain English?

3. Suppose that you wanted to study the effect of "crowding" on the weight of brown trout (*Salmo trutta*) raised in a fish hatchery in the state of Colorado in the U.S.A. Crowding is measured in terms of the number of trout raised in a container (i.e., the density of the fish in the container) and the weight of the fish is measured in grams. The trout have been raised in four containers (200, 350, 500, and 700 fish per container) for 9 months. You decide to take a random sample of trout from each of the four containers. The hypothetical data for this study are given in Fig. 8.8.

COLORADO FISH HATCHERY

Research question: What is the effect of crowding on the weight of brown trout?

WEIGHT OF BROWN TROUT (in grams)

DENSITY (number of fish per container)			
200 fish	350 fish	500 fish	700 fish
3.2	3.3	2.6	2.3
3.4	3.5	2.8	2.4
3.6	3.8	2.5	2.5
3.5	3.9	2.7	2.7
3.3	4.1	3.1	2.6
3.7	4	3.2	2.5
3.8	3.5	2.6	2.4
3.6	3.6	2.7	2.3
3.4	3.7	2.5	2.5
3.5	3.8	2.8	2.6
3.6	3.9	2.7	2.7
3.2	4.1	2.6	
3.4		2.7	
		2.9	

Fig. 8.8 Worksheet Data for Chap. 8: Practice Problem #3

(a) Enter these data on an Excel spreadsheet.
(b) Perform a *one-way ANOVA test* on these data, and show the resulting ANOVA table *underneath* the input data for the four types of crowding.
(c) If the F-value in the ANOVA table is significant, create an Excel formula to compute the ANOVA t-test comparing the average weight for 350 fish per container against the average weight for 500 fish per container, and show the results below the ANOVA table on the spreadsheet (put the standard error and the ANOVA t-test value on separate lines of your spreadsheet, and use two decimal places for each value)
(d) Print out the resulting spreadsheet so that all of the information fits onto one page
(e) Save the spreadsheet as: TROUT3

Now, write the answers to the following questions using your Excel printout:

1. What are the null hypothesis and the research hypothesis for the ANOVA F-test?
2. What is MS_b on your Excel printout?

3. What is MS_w on your Excel printout?
4. Compute $F = MS_b/MS_w$ using your calculator.
5. What is the critical value of F on your Excel printout?
6. What is the result of the ANOVA F-test?
7. What is the conclusion of the ANOVA F-test in plain English?
8. If the ANOVA F-test produced a significant difference in the average weight of the fish between the four types of crowding, what is the null hypothesis and the research hypothesis for the ANOVA t-test comparing 350 fish per container versus 500 fish per container?
9. What is the mean (average) weight for 350 fish on your Excel printout?
10. What is the mean (average) weight for 500 fish on your Excel printout?
11. What are the degrees of freedom (df) for the ANOVA t-test comparing 350 fish versus 500 fish?
12. What is the critical t value for this ANOVA t-test in Appendix E for these degrees of freedom?
13. Compute the $s.e._{ANOVA}$ using your calculator for 350 fish versus 500 fish.
14. Compute the ANOVA t-test value comparing 350 fish versus 500 fish using your calculator.
15. What is the result of the ANOVA t-test comparing 350 fish versus 500 fish?
16. What is the conclusion of the ANOVA t-test comparing 350 fish versus 500 fish in plain English?

References

Black, K. Business Statistics: For Contemporary Decision Making (6th ed.). Hoboken, NJ: John Wiley & Sons, Inc., 2010.

Gould, J. L. and Gould, G.F. Biostats Basics: A Student Handbook. New York, NY: W.H. Freeman and Company, 2002.

Weiers, R.M. Introduction to Business Statistics (7th ed.). Mason, OH: South-WesternCengage Learning, 2011.

Appendices

Appendix A: Answers to End-of-Chapter Practice Problems

T.J. Quirk et al., *Excel 2016 for Biological and Life Sciences Statistics*, Excel for Statistics, DOI 10.1007/978-3-319-39489-3

Chapter 1: Practice Problem #1 Answer (see Fig. A.1)

Number of weed seeds in a sample of grass seeds

No. of seeds		
1		
3		
2	n	17
0		
4		
6	Mean	2.94
5		
7		
0	STDEV	1.95
2		
3		
4	s.e.	0.47
2		
3		
1		
3		
4		

Fig. A.1 Answer to Chap. 1: Practice Problem #1

Chapter 1: Practice Problem #2 Answer (see Fig. A.2)

REGISTERED THREE-YEAR-OLD AYRSHIRE COWS IN MONTANA		
PERCENT OF BUTTERFAT		
3.70		
5.10		
3.95		
3.84	n	18
3.92		
4.10		
4.32	Mean	4.34
4.84		
4.61		
5.00	STDEV	0.45
4.92		
3.86		
3.97	s.e.	0.11
4.23		
4.54		
4.89		
4.23		
4.15		

Fig. A.2 Answer to Chap. 1: Practice Problem #2

Chapter 1: Practice Problem #3 Answer (see Fig. A.3)

Fairbanks, Alaska, Elementary School				
5th grade anatomy science test				
Chapter 8 (15 items)				
12				
15			n	16
13				
8				
10			MEAN	11.750
12				
13				
12			STDEV	2.910
9				
4				
11			s.e.	0.727
15				
13				
15				
12				
14				

Fig. A.3 Answer to Chap. 1: Practice Problem #3

Chapter 2: Practice Problem #1 Answer (see Fig. A.4)

Fig. A.4 Answer to Chap. 2: Practice Problem #1

FRAME NUMBERS	Duplicate frame numbers	RANDOM NO.
1	7	0.768
2	50	0.514
3	23	0.976
4	13	0.115
5	6	0.251
6	14	0.762
7	60	0.282
8	37	0.072
9	61	0.711
10	33	0.551
11	43	0.214
12	4	0.030
13	8	0.281
14	5	0.472
15	59	0.270
16	39	0.636
17	63	0.687
18	35	0.517
19	49	0.669
20	1	0.357
21	[illegible]	[illegible]
[illegible]	[illegible]	0.4.
56	19	0.187
57	45	0.421
58	21	0.127
59	38	0.429
60	17	0.110
61	11	0.729
62	42	0.837
63	54	0.610

Chapter 2: Practice Problem #2 Answer (see Fig. A.5)

Fig. A.5 Answer to Chap. 2: Practice Problem #2

FRAME NO.	Duplicate frame no.	Random number
1	58	0.678
2	50	0.292
3	43	0.526
4	42	0.386
5	86	0.394
6	24	0.719
7	22	0.028
8	11	0.113
9	104	0.504
10	105	0.661
11	61	0.117
12	41	0.169
13	79	0.263
14	93	0.405
15	85	0.037
16	54	0.298
17	16	0.960
18	77	0.756
19	15	0.270
20	112	0.569
21	28	0.868
[illegible]	10[illegible]	[illegible]
[illegible]	[illegible]	0.89[illegible]
100	96	0.130
101	48	0.722
102	108	0.724
103	109	0.735
104	33	0.118
105	3	0.903
106	90	0.209
107	110	0.141
108	62	0.391
109	88	0.820
110	60	0.734
111	98	0.029
112	94	0.632
113	59	0.521
114	67	0.634

Chapter 2: Practice Problem #3 Answer (see Fig. A.6)

Fig. A.6 Answer to Chap. 2: Practice Problem #3

FRAME NUMBERS	Duplicate frame numbers	Random number
1	58	0.230
2	7	0.544
3	37	0.221
4	49	0.732
5	26	0.905
6	65	0.634
7	48	0.185
8	63	0.131
9	15	0.485
10	25	0.157
11	21	0.260
12	36	0.835
13	43	0.452
14	11	0.782
15	10	0.409
16	39	0.593
17	72	0.257
18	59	0.696
19	16	0.846
20	54	0.901
21	52	0.498
22	3	0.010
23	45	0.174
24	[illegible]	[illegible]
[illegible]	[illegible]	[illegible]
71	60	0.862
72	9	0.936
73	20	0.762
74	18	0.910
75	5	0.080

Chapter 3: Practice Problem #1 Answer (see Fig. A.7)

MILK YIELD OF JERSEY COWS IN WISCONSIN						
	Null Hypothesis:		μ	=	68	
ANNUAL YIELD (in hundreds of pounds)						
100	Research hypothesis:		μ	≠	68	
90						
78						
70	n		19			
56						
52	Mean		60.89			
50						
55	STDEV		14.09			
58						
62	s.e.		3.23			
63						
56						
54						
52	95% confidence interval					
53						
51		Lower limit:		54.10		
56						
50		Upper limit:		67.69		
51						
	54.10 ------------------		-----60.89	------------	67.69 -----	68 --------
	lower limit		Mean		upper liimit	Ref. Value
Result:	Since the reference value is outside the confidence interval, we reject the null hypothesis and accept the research hypothesis.					
Conclusion:	The annual milk yield of the Jersey cows was significantly less this year than last year, and was probably closer to 6,100 pounds.					

Fig. A.7 Answer to Chap. 3: Practice Problem #1

Chapter 3: Practice Problem #2 Answer (see Fig. A.8)

WEIGHT OF TROUT WHEN RELEASED FROM FISH HATCHERY

Weight (g)					
254.8	Null hypothesis:		μ	=	308 g
291.2					
324.8	Research hypothesis:		μ	≠	308 g
294.0					
355.6					
347.2	n	14			
347.2					
347.2	mean	330.40			
305.2					
313.6	stdev	33.96			
366.8					
350.0	s.e.	9.08			
366.8					
361.2					

95% confidence interval

lower limit	310.79
upper limit	350.01

--------308 -------------	310.79---------	---330.40-- ------------	350.01--------
Ref. Value	lower limit	mean	upper limit

Result: Since the reference value is outside the confidence interval, we reject the null hypothesis and accept the research hypothesis.

Conclusion: The trout weighed significantly more and 308 grams when released from the fish hatchery, and their average weight was probably closer to 330 grams

Fig. A.8 Answer to Chap. 3: Practice Problem #2

Chapter 3: Practice Problem #3 Answer (see Fig. A.9)

LENGTH OF ACORNS FROM OAK TREES						
	Length (mm)					
	20	Null hypothesis:		μ	=	27
	18					
	22	Research hypothesis:		μ	≠	27
	23					
	27					
	26	n		22		
	28					
	29	Mean		29.23		
	30					
	30	STDEV		5.57		
	33					
	26	s.e.		1.19		
	27					
	28					
	38	95% confidence interval				
	37					
	30		lower limit:		26.76	
	32					
	33		upper limit:		31.70	
	31					
	37					
	38					
		26.76	-------------	27---	------ 29.23 -----	31.70
		lower limit		Ref. Value	Mean	Upper limit
		Result:	Since the reference value is inside the confidence interval, we accept the null hypothesis.			
		Conclusion:	There was no difference in the length of the acorns this year compared to five years ago.			

Fig. A.9 Answer to Chap. 3: Practice Problem #3

Chapter 4: Practice Problem #1 Answer (see Fig. A.10)

PETAL LENGTH OF A PLANT SPECIES				
Length (in cm)				
4.72				
4.54	Null hypothesis:	μ	=	3.10 cm
4.61				
4.23	Research hypothesis:	μ	≠	3.10 cm
4.12				
3.97				
3.85	n	16		
3.74				
3.65				
3.55	Mean	3.62		
3.12				
2.97				
2.86	STDEV	0.73		
2.71				
2.64				
2.56	s.e.	0.18		
	critical t	2.131		
	t-test	2.82		
Result:	Since the absolute value of 2.82 is greater than the critical t of 2.131, we reject the null hypothesis and accept the research hypothesis.			
Conclusion:	The average petal length of this plant is significantly longer this year than it was five years ago in Missouri.			

Fig. A.10 Answer to Chap. 4: Practice Problem #1

Chapter 4: Practice Problem #2 Answer (see Fig. A.11)

COLORADO RAINBOW TROUT MASS (in grams)						
	MASS (grams)					
	114		Null hypothesis:	μ	=	112 grams
	110					
	117		Research hypothesis:	μ	≠	112 grams
	112					
	115					
	116		n	15		
	112					
	125					
	118		Mean	115.60		
	113					
	120					
	112		STDEV	3.89		
	114					
	117					
	119		s.e.	1.00		
			critical t	2.145		
			t-test	3.59		
	Result:	Since the absolute value of 3.59 is greater than the critical t of 2.145, we reject the null hypothesis and accept the research hypothesis.				
	Conclusion:	The average mass of rainbow trout in southern Colorado today is significantly more than it was five years ago.				

Fig. A.11 Answer to Chap. 4: Practice Problem #2

Chapter 4: Practice Problem #3 Answer (see Fig. A.12)

TREE FROG MATING CALL DURATION						
	DURATION (in msec)		Null hypothesis:	μ	=	160 msec
	100					
	120		Research hypothesis:	μ	≠	160 msec
	135					
	147					
	180		n	18		
	180					
	220		Mean	230.78		
	295					
	287					
	289		STDEV	68.62		
	300					
	295					
	276		s.e.	16.17		
	268					
	254					
	264		critical t	2.110		
	248					
	296					
			t-test	4.38		
	Result:	Since the absolute value of 4.38 is greater than the critical t of 2.110, we reject the null hypothesis and accept the research hypothesis.				
	Conclusion:	The mating call note duration of the tree frog (*Hyla versicolor*) is significantly longer today than it was eight years ago.				

Fig. A.12 Answer to Chap. 4: Practice Problem #3

Chapter 5: Practice Problem #1 Answer (see Fig. A.13)

WING LENGTH (mm) OF A SPECIES OF MOSQUITOES IN TWO REGIONS

Group	n	Mean	STDEV
1 North	124	3.2	1.2
2 South	135	3.4	1.3

Null hypothesis:	μ_1	=	μ_2
Research hypothesis:	μ_1	≠	μ_2

STDEV1 squared / n1	0.012
STDEV2 squared / n2	0.013
E13 + E16	0.024
s.e.	0.155
critical t	1.96
t-test	-1.287

Result: Since the absolute value of − 1.287 is less than the critical t value of 1.96, we accept the null hypothesis.

Conclusion: There was no difference in wing length of the species of mosquitoes in the North vs. the South regions.

Fig. A.13 Answer to Chap. 5: Practice Problem #1

Chapter 5: Practice Problem #2 Answer (see Fig. A.14)

BODY MASS OF HOUSE FINCHES (in g)

EXPERIMENTAL GROUP	CONTROL GROUP
19.9	19.8
21.6	19.9
20.1	20.1
21.4	20.5
22.1	20.3
21.1	20.8
21.6	21.1
21.4	21.3
21.8	21.6
22.1	21.8
22.2	22.0
22.6	
22.4	

Group	n	Mean	STDEV
1 Experimental group	13	21.56	0.82
2 Control group	11	20.84	0.78

Null hypothesis:	μ_1	=	μ_2
Research hypothesis:	μ_1	≠	μ_2

1/n1 + 1/n2	0.17
(n1 - 1) x S1 squared	7.99
(n2 - 1) x S2 squared	6.05
n1 + n2 - 2	22
s.e.	0.33
critical t	2.074
t-test	2.22

Result: Since the absolute value of 2.22 is greater than the critical t of 2.074, we reject the null hypothesis and accept the research hypothesis.

Conclusion: The body mass of house finches in the experimental group that was fed a special diet was significantly greater than the body mass of the control group that received the regular diet (21.56 g vs. 20.84 g).

Fig. A.14 Answer to Chap. 5: Practice Problem #2

Chapter 5: Practice Problem #3 Answer (see Fig. A.15)

GPA OF BIOLOGY MAJORS WHO HAVE COMPLETED ALL BIOLOGY REQUIRED COURSES

MEN	WOMEN
2.45	2.83
2.53	2.74
2.64	2.86
2.72	3.32
2.85	3.36
2.96	3.64
3.01	3.56
3.11	3.56
3.24	3.64
3.35	3.37
3.36	3.67
3.38	3.91
3.21	3.92
3.52	3.64
3.64	3.71
3.75	
3.86	

Group	n	Mean	STDEV
1 Men	17	3.15	0.42
2 Women	15	3.45	0.37

Null hypothesis:	μ_1	=	μ_2
Research hypothesis:	μ_1	≠	μ_2

(n1 - 1) x STDEV1 squared	2.86
(n2 - 1) x STDEV2 squared	1.95
n1 + n2 – 2	30
1/n1 + 1/n2	0.13
s.e.	0.14
critical t	2.042
t-test	-2.09

Result: Since the absolute value of – 2.09 is greater than the critical t of 2.042, we reject the null hypothesis and accept the research hypothesis.

Conclusion: Female biology majors who have completed all of the Biology required courses for majors had significantly higher GPAs than male biology majors (3.45 vs. 3.15).

Fig. A.15 Answer to Chap. 5: Practice Problem #3

Chapter 6: Practice Problem #1 Answer (see Fig. A.16)

What is the relationship between the age of trees and the number of growth rings on the trees?

AGE (Years)	NUMBER OF GROWTH RINGS
1.0	1.1
5.1	3.5
10.2	8.0
15.3	6.5
20.5	22.0
25.2	17.0
30.1	26.3
34.8	23.6
correlation	0.93

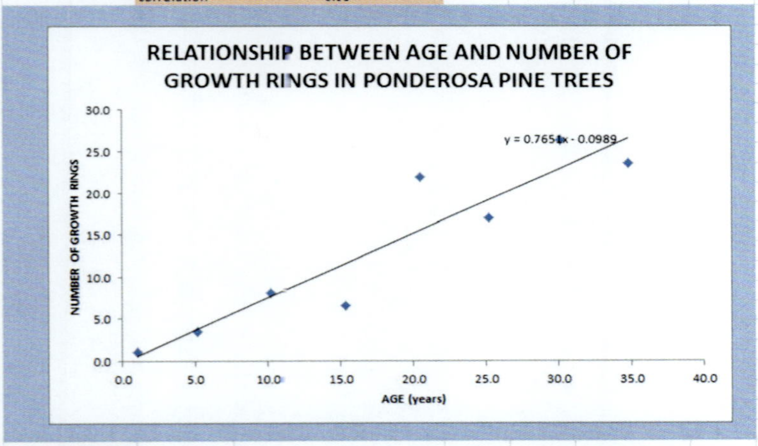

SUMMARY OUTPUT

Regression Statistics	
Multiple R	0.93
R Square	0.865
Adjusted R Square	0.842
Standard Error	3.922
Observations	8

ANOVA

	df	*SS*	*MS*	*F*	*Significance F*
Regression	1	591.089	591.089	38.436	0.0008
Residual	6	92.271	15.378		
Total	7	683.360			

	Coefficients	*Standard Error*	*t Stat*	*P-value*	*Lower 95%*
Intercept	-0.099	2.595	-0.038	0.971	-6.448
X Variable 1	0.765	0.123	6.200	0.001	0.463

Fig. A.16 Answer to Chap. 6: Practice Problem #1

Chapter 6: Practice Problem #1 (continued)

1. $r = +.93$
2. $a = \text{y-intercept} = -0.099$
3. $b = \text{slope} = 0.765$
4. $Y = a + bX$
   $Y = -0.099 + 0.765X$
5. $Y = -0.099 + 0.765(20)$
   $Y = -0.099 + 15.3$
   $Y = 15.2$ growth rings

Chapter 6: Practice Problem #2 Answer (see Fig. A.17)

BODY WEIGHT VS. GILL WEIGHT IN *Pachygrapsus marmoratus*

BODY WEIGHT IN GRAMS (g)	GILL WEIGHT IN MILLIGRAMS (mg)
3.2	40.2
10.5	90.5
15.6	120.6
17.4	320.7
19.8	240.9
5.3	70.5
13.5	230.7
22.1	370.5
20.4	280.6
18.3	230.4
8.4	165.7

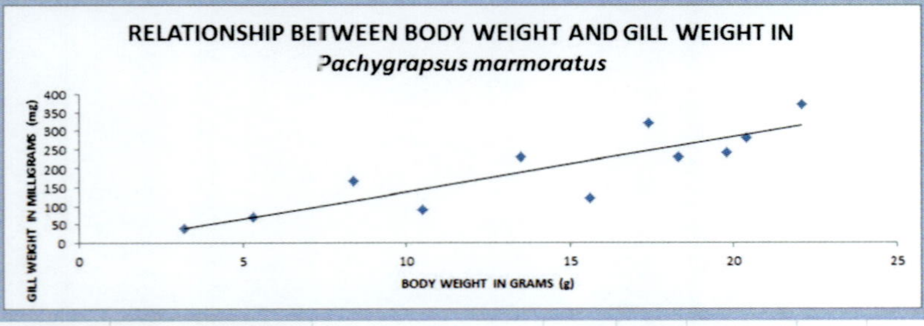

SUMMARY OUTPUT

Regression Statistics	
Multiple R	0.87
R Square	0.754
Adjusted R Square	0.726
Standard Error	56.168
Observations	11

ANOVA

	df	*SS*	*MS*	*F*	*Significance F*
Regression	1	86922.516	86922.516	27.552	0.0005
Residual	9	28393.280	3154.809		
Total	10	115315.796			

	Coefficients	*Standard Error*	*t Stat*	*P-value*	*Lower 95%*
Intercept	-8.28	42.527	-0.195	0.850	-104.481
X Variable 1	14.58	2.777	5.249	0.001	8.296

Fig. A.17 Answer to Chap. 6: Practice Problem #2

Chapter 6: Practice Problem #2 (continued)

(2b) about 290 mg

1. $r = +.87$
2. $a = \text{y-intercept} = -8.28$
3. $b = \text{slope} = 14.58$
4. $Y = a + bX$
   $Y = -8.28 + 14.58X$
5. $Y = -8.28 + 14.58(15)$
   $Y = -8.28 + 218.7$
   $Y = 210.42\,\text{mg}$

Chapter 6: Practice Problem #3 Answer (see Fig. A.18)

Is there a relationship between body length and forewing length in mayflies?

Body Length (mm)	Forewing Length (mm)
5.60	4.20
6.20	5.10
5.80	4.60
6.40	5.20
7.10	5.90
7.20	6.00
6.20	5.10
6.50	5.20
6.80	5.30
6.90	5.70
5.60	4.30
correlation	0.98

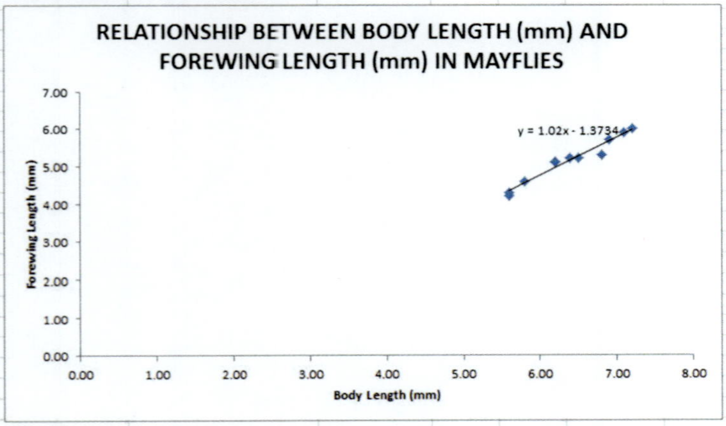

SUMMARY OUTPUT

Regression Statistics	
Multiple R	0.98
R Square	0.959
Adjusted R Square	0.954
Standard Error	0.127
Observations	11

ANOVA

	df	*SS*	*MS*	*F*	*Significance F*
Regression	1	3.401	3.401	209.724	1.53031E-07
Residual	9	0.146	0.016		
Total	10	3.547			

	Coefficients	*Standard Error*	*t Stat*	*P-value*	*Lower 95%*
Intercept	-1.373	0.452	-3.040	0.014	-2.395
X Variable 1	1.020	0.070	14.482	0.000	0.861

Fig. A.18 Answer to Chap. 6: Practice Problem #3

Chapter 6: Practice Problem #3 (continued)

1. $r = +.98$
2. $a = \text{y-intercept} = -1.373$
3. $b = \text{slope} = 1.020$
4. $Y = a + bX$
   $Y = -1.373 + 1.020X$
5. $Y = -1.373 + 1.020(6.50)$
   $Y = -1.373 + 6.630$
   $Y = 5.26$ mm

Chapter 7: Practice Problem #1 Answer (see Fig. A.19)

SEED PRODUCTION

How well do water, light, fertilizer, and space predict the number of seeds produced per pod?

AVERAGE SEEDS PER POD	WATER (ml)	LIGHT RECEIVED (min)	SPACE (cm)	FERTILIZER (μl)
3.25	600	620	12.5	650
3.42	520	550	10	600
2.85	510	540	5	500
2.65	480	460	2.5	510
3.65	720	710	15	630
3.16	570	610	7.5	550
3.56	710	650	10	610
2.35	500	480	5	430
2.86	450	470	7.5	450
2.95	560	530	10	550
3.15	550	580	10	580
3.45	610	620	12.5	620

SUMMARY OUTPUT

Regression Statistics	
Multiple R	0.93
R Square	0.857875018
Adjusted R Square	0.776660743
Standard Error	0.184379009
Observations	12

ANOVA

	df	*SS*	*MS*	*F*	*Significance F*
Regression	4	[illegible]436397334	0.359099334	10.56310622	0.004331962
Residual	7	[illegible]237969332	0.033995619		
Total	11	[illegible]674366667			

	Coefficients	*Standard Error*	*t Stat*	*P-value*	*Lower 95%*
Intercept	0.5682	[illegible]632835292	0.897786192	0.399124683	-0.928266892
WATER (ml)	-0.0004	[illegible]00181073	-0.210590788	0.83920554	-0.00466302
LIGHT RECEIVED (min)	0.0022	[illegible]002421544	0.907419349	0.39434945	-0.003528685
SPACE (cm)	0.0200	[illegible]029379257	0.682445079	0.516897071	-0.049421175
FERTILIZER (μl)	0.0024	[illegible]001545116	1.543035862	0.166730022	-0.001269449

	AVERAGE SEEDS PER POD	*WATER (ml)*	*LIGHT RECEIVED (min)*	*SPACE (cm)*	*FERTILIZER (μl)*
AVERAGE SEEDS PER POD	1				
WATER (ml)	0.79	1			
LIGHT RECEIVED (min)	0.87	0.93	1		
SPACE (cm)	0.83	0.74	0.82	1	
FERTILIZER (μl)	0.89	0.76	0.83	0.81	1

Fig. A.19 Answer to Chap. 7: Practice Problem #1

Chapter 7: Practice Problem #1 (continued)

1. Multiple correlation = .93
2. y-intercept = 0.5682
3. $b_1 = -0.0004$
4. $b_2 = 0.0022$
5. $b_3 = 0.0200$
6. $b_4 = 0.0024$
7. $Y = a + b_1X_1 + b_2X_2 + b_3X_3 + b_4X_4$
   $Y = 0.5682 - 0.0004X_1 + 0.0022X_2 + 0.0200X_3 + 0.0024X_4$
8. $Y = 0.5682 - 0.0004(610) + 0.0022(550) + 0.0200(3) + 0.0024(610)$
   $Y = 0.5682 - 0.244 + 1.21 + 0.06 + 1.464$
   $Y = 3.302 - 0.244$
   $Y = 3.06$
9. 0.79
10. 0.87
11. 0.83
12. 0.89
13. 0.74
14. 0.83
15. The best predictor of AVERAGE SEEDS PER POD was FERTILIZER ($r = +0.89$).
16. The four predictors combined predict the AVERAGE SEEDS PER POD at $R_{xy} = .93$, and this is much better than the best single predictor by itself.

Chapter 7: Practice Problem #2 Answer (see Fig. A.20)

	PLANT-GROWTH DATA	
PRODUCTIVITY (g/m squared/year)	MEAN ANNUAL PRECIPITATION (cm/year)	MEAN ANNUAL TEMPERATURE (degrees C)
250	100	-13
300	300	2.5
400	100	1.5
350	200	2.5
450	300	0
550	400	-2.5
500	200	-4
650	200	-6.5
600	400	-2
750	200	4.5
800	400	-9.5
850	500	2
950	200	3
1000	500	3

SUMMARY OUTPUT

Regression Statistics	
Multiple R	0.65
R Square	0.417
Adjusted R Square	0.311
Standard Error	200.688
Observations	14

ANOVA

	df	*SS*	*MS*
Regression	2	316965.7145	158482.8572
Residual	11	443034.2855	40275.84414
Total	13	760000	

	Coefficients	*Standard Error*	*t Stat*
Intercept	322.246	128.88	2.50
MEAN ANNUAL PRECIPITATION (cm/year)	1.040	0.41	2.53
MEAN ANNUAL TEMPERATURE (degrees C)	9.005	10.68	0.84

	PRODUCTIVITY (g/m squared/year)	*MEAN ANNUAL PRECIPITATION (cm/year)*	*MEAN ANNUAL TEMPERATURE (degrees C)*
PRODUCTIVITY (g/m squared/year)	1		
MEAN ANNUAL PRECIPITATION (cm/year)	0.62	1	
MEAN ANNUAL TEMPERATURE (degrees C)	0.28	0.14	1

Fig. A.20 Answer to Chap. 7: Practice Problem #2

Chapter 7: Practice Problem #2 (continued)

1. $R_{xy} = 0.65$
2. a = y-intercept = 322.246
3. $b_1 = 1.04$
4. $b_2 = 9.005$
5. $Y = a + b_1X_1 + b_2X_2$
   $Y = 322.246 + 1.04\ X_1 + 9.005\ X_2$
6. $Y = 322.246 + 1.04(300) + 9.005(2)$
   $Y = 322.246 + 312 + 18.01$
   $Y = 652$ g/meter squared/year
7. +0.62
8. +0.28
9. +0.14
10. Mean annual precipitation is the better predictor of productivity ($r = +.62$)
11. The two predictors combined predict productivity slightly better ($R_{xy} = .65$) than the better single predictor by itself

Chapter 7: Practice Problem #3 Answer (see Fig. A.21)

FROSH GPA	SAT-VERBAL	SAT-MATH	HS GPA	AP BIOLOGY TEST
3.23	650	510	3.55	2
2.90	490	420	2.96	2
2.80	630	540	3.27	5
3.42	520	520	3.45	4
2.80	560	540	2.90	3
2.90	410	420	2.80	2
2.35	450	460	2.63	2
2.58	420	430	2.71	1
3.12	560	520	3.26	3
3.47	650	680	3.45	4

SUMMARY OUTPUT

Regression Statistics	
Multiple R	0.95
R Square	0.911
Adjusted R Square	0.839
Standard Error	0.143
Observations	10

ANOVA

	df	*SS*	*MS*	*F*	*Significance F*
Regression	4	1.043	0.261	12.719	0.008
Residual	5	0.102	0.020		
Total	9	1.145			

	Coefficients	*Standard Error*	*t Stat*	*P-value*	*Lower 95%*
Intercept	-0.684	0.545	-1.254	0.265	-2.085
SAT-VERBAL	-0.003	0.001	-2.598	0.048	-0.006
SAT-MATH	0.002	0.001	2.029	0.098	-0.001
HS GPA	1.378	0.259	5.320	0.003	0.712
AP BIOLOGY TEST	-0.048	0.057	-0.845	0.437	-0.195

	FROSH GPA	*SAT-VERBAL*	*SAT-MATH*	*HS GPA*	*AP BIOLOGY TEST*
FROSH GPA	1				
SAT-VERBAL	0.59	1			
SAT-MATH	0.61	0.80	1		
HS GPA	0.88	0.83	0.66	1	
AP BIOLOGY TEST	0.47	0.65	0.71	0.61	1

Fig. A.21 Answer to Chap. 7: Practice Problem #3

Chapter 7: Practice Problem #3 (continued)

1. Multiple correlation = +.95
2. a = y-intercept = −0.684
3. $b_1 = -0.003$
4. $b_2 = 0.002$
5. $b_3 = 1.378$
6. $b_4 = -0.048$
7. $Y = a + b_1X_1 + b_2X_2 + b_3X_3 + b_4X_4$
   $Y = -0.684 - 0.003\ X_1 + 0.002\ X_2 + 1.378\ X_3 - 0.048\ X_4$
8. $Y = -0.684 - 0.003(650) + 0.002(630) + 1.378(3.47) - 0.048(4)$
   $Y = -0.684 - 1.95 + 1.26 + 4.78 - 0.19$
   $Y = 6.04 - 2.824$
   $Y = 3.22$
9. +0.59
10. +0.61
11. +0.88
12. +0.47
13. +0.80
14. +0.61
15. +0.71
16. +.65
17. The best single predictor of FROSH GPA was HS GPA (r = .88).
18. The four predictors combined predict FROSH GPA at $R_{xy} = .95$, and this is much better than the best single predictor by itself.

Chapter 8: Practice Problem #1 Answer (see Fig. A.22)

DAISY FLEABANE GROWTH IN ELEVATED NITROGEN SOILS

No Nitrogen	Low Nitrogen	High Nitrogen
500	550	800
550	600	1000
700	750	900
650	700	1100
500	600	1400
550	650	1600
600	600	1500
650	700	1800
600	650	1500
	750	1800
		2000
		1600

Anova: Single Factor

SUMMARY

Groups	*Count*	*Sum*	*Average*	*Variance*
No Nitrogen	9	5300	588.89	4861.11
Low Nitrogen	10	6550	655.00	4694.44
High Nitrogen	12	17000	1416.67	148787.88

ANOVA

Source of Variation	*SS*	*df*	*MS*	*F*	*P-value*	*F crit*
Between Groups	4645581.54	2	2322790.77	37.86	1.09E-08	3.34
Within Groups	1717805.56	28	61350.20			
Total	6363387.10	30				

Low Nitrogen vs. High Nitrogen

1/10 + 1/12	0.18
s.e. ANOVA	106.05
ANOVA t-test	-7.18

Fig. A.22 Answer to Chap. 8: Practice Problem #1

Chapter 8: Practice Problem #1 (continued)

Let Group 1 = No Nitrogen, Group 2 = Low Nitrogen, and Group 3 = High Nitrogen

1. $H_0 : \mu_1 = \mu_2 = \mu_3$
   $H_1 : \mu_1 \neq \mu_2 \neq \mu_3$
2. $MS_b = 2{,}322{,}790.77$
3. $MS_w = 61{,}350.20$
4. $F = 2{,}322{,}790/61{,}350 = 37.86$
5. critical $F = 3.34$
6. Result: Since 37.86 is greater than 3.34, we reject the null hypothesis and accept the research hypothesis
7. There was a significant difference in the weight of the Daisy Fleabane between the three treatments.

TREATMENT 2 vs. TREATMENT 3

8. $H_0 : \mu_2 = \mu_3$
   $H_1 : \mu_2 \neq \mu_3$
9. 655
10. 1416.67
11. $df = 31 - 3 = 28$
12. critical $t = 2.048$
13. $1/10 + 1/12 = 0.10 + 0.08 = 0.18$

$$\text{s.e.} = \text{SQRT}(61{,}350.20{*}0.18) = \text{SQRT}(11{,}043.04) = 105.086$$

14. ANOVA $t = (655 - 1416.67)/105.086 = -761.67/105.086 = -7.25$
15. Result: Since the absolute value of -7.25 is greater than 2.048, we reject the null hypothesis and accept the research hypothesis
16. Conclusion: Daisy Fleabane flowers weighed significantly more when the soil contained high Nitrogen than when the soil contained low Nitrogen (1417 mg vs. 655 mg).

Chapter 8: Practice Problem #2 Answer (see Fig. A.23)

INTRODUCTORY BIOLOGY: REASONS FOR TAKING THIS COURSE

Cumulative GPAs when entering this course

A	B	C	D
Premed required course	Biology major	Other Science major	General Ed requirement
2.84	2.60	2.50	2.00
3.26	2.86	2.65	2.10
3.05	2.95	2.70	2.25
3.15	3.00	2.55	2.35
3.35	3.14	2.80	2.43
3.40	3.25	3.20	2.67
3.55	3.35	3.35	3.00
3.68	3.60	3.50	3.16
3.84	3.70	3.60	3.26
3.90		3.70	3.55
		3.8	3.35
			3.25

Anova: Single Factor

SUMMARY

Groups	*Count*	*Sum*	*Average*	*Variance*
Premed required course	10	34.02	3.40	0.12
Biology major	9	28.45	3.16	0.13
Other Science major	11	34.35	3.12	0.24
General Ed requirement	12	33.37	2.78	0.29

ANOVA

Source of Variation	*SS*	*df*	*MS*	*F*	*P-value*	*F crit*
Between Groups	2.17	3	0.72	3.56	0.02	2.85
Within Groups	7.74	38	0.20			
Total	9.92	41				

Premed vs. Other Science major

1/10+1/11	0.19
s.e. ANOVA	0.20
ANOVA t-test	1.42

Fig. A.23 Answer to Chap. 8: Practice Problem #2

Chapter 8: Practice Problem #2 (continued)

1. Null hypothesis : $\mu_A = \mu_B = \mu_C = \mu_D$
   Research hypothesis : $\mu_A \neq \mu_B \neq \mu_C \neq \mu_D$
2. $MS_b = 0.72$
3. $MS_w = 0.20$
4. $F = 0.72/0.20 = 3.60$
5. critical $F = 2.85$
6. Since the F-value of 3.60 is greater than the critical F value of 2.85, we reject the null hypothesis and accept the research hypothesis.
7. There was a significant difference in cumulative GPA between the four groups of students in their reasons for taking Introductory Biology.
8. Null hypothesis : $\mu_A = \mu_C$
   Research hypothesis : $\mu_A \neq \mu_C$
9. 3.40
10. 3.12
11. degrees of freedom $= 42 - 4 = 38$
12. critical $t = 2.024$
13. $s.e._{ANOVA} = SQRT(MS_w \times \{1/10 + 1/11\}) = SQRT(0.20 \times 0.19) =$ $SQRT(0.038) = 0.20$
14. ANOVA $t = (3.40 - 3.12)/0.20 = 1.42$
15. Since the absolute value of 1.42 is LESS than the critical t of 2.024, we accept the null hypothesis.
16. There was no difference in cumulative GPA between Premed students and Other Science majors in their reasons for taking the Introductory Biology course.

Chapter 8: Practice Problem #3 Answer (see Fig. A.24)

COLORADO FISH HATCHERY

Research question: What is the effect of crowding on the weight of brown trout?

WEIGHT OF BROWN TROUT (in grams)

DENSITY (number of fish per container)			
200 fish	350 fish	500 fish	700 fish
3.2	3.3	2.6	2.3
3.4	3.5	2.8	2.4
3.6	3.8	2.5	2.5
3.5	3.9	2.7	2.7
3.3	4.1	3.1	2.6
3.7	4	3.2	2.5
3.8	3.5	2.6	2.4
3.6	3.6	2.7	2.3
3.4	3.7	2.5	2.5
3.5	3.8	2.8	2.6
3.6	3.9	2.7	2.7
3.2	4.1	2.6	
3.4		2.7	
		2.9	

Anova: Single Factor

SUMMARY

Groups	*Count*	*Sum*	*Average*	*Variance*
200 fish	13	45.2	3.48	0.03
350 fish	12	45.2	3.77	0.06
500 fish	14	38.4	2.74	0.04
700 fish	11	27.5	2.50	0.02

ANOVA

Source of Variation	*SS*	*df*	*MS*	*F*	*P-value*	*F crit*
Between Groups	12.89	3	4.30	106.05	1.11E-20	2.81
Within Groups	1.86	46	0.04			
Total	14.76	49				

CROWDING: 350 fish vs. 500 fish per container

1/n 350 + 1/n 500	0.15
s.e of 350 fish vs. 500 fish	0.08
ANOVA t	12.93

Fig. A.24 Answer to Chap. 8: Practice Problem #3

Chapter 8: Practice Problem #3 (continued)

Let 200 fish = Group 1, 350 fish = Group 2, 500 fish = Group 3, and 700 fish = Group 4

1. Null hypothesis : $\mu_1 = \mu_2 = \mu_3 = \mu_4$
   Research hypothesis : $\mu_1 \neq \mu_2 \neq \mu_3 \neq \mu_4$
2. $MS_b = 4.30$
3. $MS_w = 0.04$
4. $F = 4.30/0.04 = 107.50$
5. critical $F = 2.81$
6. Result: Since the F-value of 107.50 is greater than the critical F value of 2.81, we reject the null hypothesis and accept the research hypothesis.
7. Conclusion: There was a significant difference between the four types of crowding conditions in the weight of the brown trout.
8. Null hypothesis : $\mu_2 = \mu_3$
   Research hypothesis : $\mu_2 \neq \mu_3$
9. 3.77 grams
10. 2.74 grams
11. degrees of freedom = 50 − 4 = 46
12. critical t = 1.96
13. $s.e._{ANOVA} = SQRT(MS_w \times \{1/12 + 1/14\}) = SQRT(0.04 \times 0.15) =$ $SQRT(0.006) = 0.08$
14. ANOVA t = (3.77 − 2.74)/0.08 = 12.88
15. Since the absolute value of 12.88 is greater than the critical t of 1.96, we reject the null hypothesis and accept the research hypothesis.
16. Brown trout raised in a container with 350 fish weighed significantly more than brown trout raised in a container with 500 fish (3.77 grams vs. 2.74 grams).

Appendix B: Practice Test

Chapter 1: Practice Test

Suppose that you were hired as a research assistant on a project involving hummingbirds, and that your responsibility on this team was to measure the width (at the widest part) of hummingbird eggs (measured in millimeters). You want to try out your Excel skills on a small sample of eggs and measure them. The hypothetical data are given below (see Fig. B.1).

Fig. B.1 Worksheet Data for Chap. 1 Practice Test (Practical Example)

WIDTH (at the widest part) OF HUMMINGBIRD EGGS

WIDTH (mm)
9.4
9.6
9.7
9.8
10.1
9.5
9.4
10.2
9.6
10.3
9.6
9.8
10.1
9.4
9.5
9.7

(a) Create an Excel table for these data, and then use Excel to the right of the table to find the sample size, mean, standard deviation, and standard error of the mean for these data. Label your answers, and round off the mean, standard deviation, and standard error of the mean to two decimal places.

(b) Save the file as: eggs3

Chapter 2: Practice Test

Suppose that you were hired to test the fluoride levels in drinking water in Jefferson County, Colorado. Historically, there are a total of 42 water sample collection sites. Because of budget constraints, you need to choose a random sample of 12 of these 42 water sample collection sites.

(a) Set up a spreadsheet of frame numbers for these water samples with the heading: FRAME NUMBERS
(b) Then, create a separate column to the right of these frame numbers which duplicates these frame numbers with the title: Duplicate frame numbers.
(c) Then, create a separate column to the right of these duplicate frame numbers called RAND NO. and use the *=RAND()* function to assign random numbers to all of the frame numbers in the duplicate frame numbers column, and change this column format so that three decimal places appear for each random number.
(d) Sort the *duplicate frame numbers and random numbers* into a random order.
(e) Print the result so that the spreadsheet fits onto one page.
(f) Circle on your printout the I.D. number of the first 12 water sample locations that you would use in your test.
(g) Save the file as: RAND15

Important note: Note that everyone who does this problem will generate a different random order of water sample sites ID numbers since Excel assign a different random number each time the RAND() command is used. For this reason, the answer to this problem given in this Excel Guide will have a completely different sequence of random numbers from the random sequence that you generate. This is normal and what is to be expected.

Chapter 3: Practice Test

Suppose that you have been asked to analyze some environmental impact data from the state of Texas in terms of the amount of SO_2 concentration in the atmosphere in different sites of Texas compared to 3 years ago to see if this concentration (and the air the people who live there breathe) has changed. SO_2 is measured in parts per billion (ppb). Three years ago, when this research was last done, the average concentration of SO_2 in these sites was 120 ppb. Since then, the state has undertaken a comprehensive program to improve the air that people in these sites breathe, and you have been asked to "run the data" to see if any change has occurred.

Is the air that people breathe in these sites now different from the air that people breathed 3 years ago? You have decided to test your Excel skills on a sample of hypothetical data given in Fig. B.2

CONCENTRATION OF SO_2 IN THE ATMOSPHERE IN PARTS PER BILLION (ppb)
ppb
390
332
186
85
29
135
86
54
28
35
37
28
18
32
24
19
21
35
31
18
20
21
18

Fig. B.2 Worksheet Data for Chap. 3 Practice Test (Practical Example)

(a) Create an Excel table for these data, and use Excel to the right of the table to find the sample size, mean, standard deviation, and standard error of the mean for these data. Label your answers, and round off the mean, standard deviation, and standard error of the mean to two decimal places in number format.
(b) By hand, write the null hypothesis and the research hypothesis on your printout.
(c) Use Excel's *TINV function* to find the 95 % confidence interval about the mean for these data. Label your answers. Use two decimal places for the confidence interval figures in number format.
(d) On your printout, draw a diagram of this 95 % confidence interval by hand, including the reference value.
(e) On your spreadsheet, enter the *result.*
(f) On your spreadsheet, enter the *conclusion in plain English.*
(g) Print the data and the results so that your spreadsheet fits onto one page.
(h) Save the file as: PARTS3

Chapter 4: Practice Test

Suppose that you wanted to measure the evolution of birds after a severe environmental change. Specifically, you want to study the effect of a severe drought with very little rainfall on the beak depth of finches on a small island in the Pacific. Let's suppose that the drought has lasted 6 years and that before the drought, the average beak depth of finches on this island was 9.2 millimeters (mm). Has the beak depth of finches changed after this environmental challenge? You measure the beak depth of a sample of finches, and the hypothetical data is given in Fig. B.3.

BEAK DEPTH OF FINCHES
BEAK DEPTH (mm)
6.1
6.8
10.3
10.8
7.5
9.8
9.6
7.2
7.8
9.9
7.6
9.7
11.4
8.3
11.3
11.4
11.3
10.8
10.3
10.6
10.8
10.9
11.1
11.3
11.2
11.4
11.2
11.3
11.1

Fig. B.3 Worksheet Data for Chap. 4 Practice Test (Practical Example)

(a) Write the null hypothesis and the research hypothesis on your spreadsheet.
(b) Create a spreadsheet for these data, and then use Excel to find the sample size, mean, standard deviation, and standard error of the mean to the right of the data set. Use number format (three decimal places) for the mean, standard deviation, and standard error of the mean.
(c) Type the *critical t* from the t-table in Appendix E onto your spreadsheet, and label it.
(d) Use Excel to compute the t-test value for these data (use three decimal places) and label it on your spreadsheet.
(e) Type the *result* on your spreadsheet, and then type the *conclusion in plain English* on your spreadsheet.

(f) Save the file as:
beak3

Chapter 5: Practice Test

Suppose that you wanted to study the duration of hibernation of a species of hedgehogs (*Erinaceus europaeus*) in two regions of the United States (NORTH vs. SOUTH). Suppose, further, that researchers have captured hedgehogs in these regions, attached radio tags to their bodies, and then released them back into the site where they were captured. The researchers monitored the movements of the hedgehogs during the winter months to determine the number of days that they did not leave their nests. The researchers have selected a random sample of hedgehogs from each region, and you want to test your Excel skills on the hypothetical data given in Fig. B.4.

Fig. B.4 Worksheet Data for Chap. 5 Practice Test (Practical Example)

NUMBER OF DAYS SPENT IN THE NEST	
NORTH REGION	SOUTH REGION
95	90
98	92
99	94
112	93
103	95
105	98
106	100
109	102
110	103
111	104
112	98
114	102
95	99
98	
103	

(a) Write the null hypothesis and the research hypothesis.
(b) Create an Excel table that summarizes these data.
(c) Use Excel to find the standard error of the difference of the means.
(d) Use Excel to perform a *two-group t-test*. What is the value of *t* that you obtain (use two decimal places)?
(e) On your spreadsheet, type the *critical value of t* using the t-table in Appendix E.

(f) Type the *result* of the test on your spreadsheet.
(g) Type your *conclusion in plain English* on your spreadsheet.
(h) Save the file as: HEDGE3
(i) Print the final spreadsheet so that it fits onto one page.

Chapter 6: Practice Test

How does elevation effect overall plant height? Suppose that you wanted to study this question using the height of yarrow plants. The plants were germinated from seeds collected at different elevations from 15 different sites in the western United States. The plants were reared in a greenhouse to control for the effect of temperature on plant height. The hypothetical data are given in Fig. B.5.

Fig. B.5 Worksheet Data for Chap. 6 Practice Test (Practical Example)

SITE	ELEVATION (m)	HEIGHT (cm)
1	1200	75
2	1350	65
3	1400	68
4	1550	60
5	1600	45
6	1750	50
7	1850	46
8	1900	48
9	2000	45
10	2100	25
11	2250	23
12	2300	20
13	2400	21
14	2850	18
15	3000	15

Create an Excel spreadsheet, and enter the data.

(a) create an *XY scatterplot* of these two sets of data such that:

- top title: RELATIONSHIP BETWEEN ELEVATION AND HEIGHT OF YARROW PLANTS
- x-axis title: Elevation (m)
- y-axis title: Height (cm)
- move the chart below the table
- re-size the chart so that it is 7 columns wide and 25 rows long
- delete the legend
- delete the gridlines

(b) Create the *least-squares regression line* for these data on the scatterplot.

(c) Use Excel to run the regression statistics to find the *equation for the least-squares regression line* for these data and display the results below the chart on your spreadsheet. Add the regression equation to the chart. Use number format (three decimal places) for the correlation and for the coefficients

Print *just the input data and the chart* so that this information fits onto one page in portrait format.

Then, print *just the regression output table* on a separate page so that it fits onto that separate page in portrait format.

By hand:

(d) Circle and label the value of the *y-intercept* and the *slope* of the regression line on your printout.

(e) Write the regression equation *by hand* on your printout for these data (use three decimal places for the y-intercept and the slope).

(f) Circle and label the *correlation* between the two sets of scores in the regression analysis summary output table on your printout.

(g) Underneath the regression equation you wrote by hand on your printout, use the regression equation to predict the average height of the yarrow plant you would predict for an elevation of 2000 meters.

(h) *Read from the graph*, the average height of the yarrow plant you would predict for an elevation of 2500 meters, and write your answer in the space immediately below:

____

(i) save the file as: Yarrow3

Chapter 7: Practice Test

Suppose that you have been hired by the United States Department of Agriculture (USDA) to analyze corn yields from Iowa farms over 1 year (a single growing season). Suppose, further, that these data will represent a pilot study that will be included in a larger ongoing analysis of corn yield in the Midwest. You want to determine if you can predict the amount of corn produced in bushels per acre (bu/acre) based on three predictors: (1) water measured in inches of rainfall per year (in/yr), (2) fertilizer measured in the amount of nitrogen applied to the soil in pounds per acre (lbs/acre), and (3) average temperature during the growing season measured in degrees Fahrenheit (°F). To check your skills in Excel, you have selected a random sample of corn from each of eleven farms selected randomly and recorded the hypothetical given in Fig. B.6.

CORN YIELDS ON IOWA FARMS			
YIELD (bu/acre)	RAINFALL (in/yr)	NITROGEN (lbs/acre)	TEMPERATURE (degrees F)
240	28.8	190	80
255	29.5	195	79
299	27.6	230	79
230	28.3	175	78
300	31.5	250	79
180	27.1	250	82
294	28.9	220	78
278	27.1	200	74
284	28.8	204	80
194	30.7	185	81
170	27.9	190	88

Fig. B.6 Worksheet Data for Chap. 7 Practice Test (Practical Example)

(a) create an Excel spreadsheet using YIELD as the criterion (Y), and the other variables as the three predictors of this criterion (X_1 = RAINFALL, X_2 = NITROGEN, and X_3 = TEMPERATURE).
(b) Use Excel's *multiple regression* function to find the relationship between these four variables and place the SUMMARY OUTPUT below the table.
(c) Use number format (two decimal places) for the multiple correlation on the Summary Output, and use two decimal places for the coefficients in the SUMMARY OUTPUT.
(d) Save the file as: yield15
(e) Print the table and regression results below the table so that they fit onto one page.

Answer the following questions using your Excel printout:

1. What is the multiple correlation R_{xy}?
2. What is the y-intercept a?
3. What is the coefficient for RAINFALL b_1?
4. What is the coefficient for NITROGEN b_2?
5. What is the coefficient for TEMPERATURE b_3?
6. What is the multiple regression equation?
7. Predict the corn yield you would expect for rainfall of 28 inches per year, nitrogen at 205 pounds/acre, and temperature of 83 degrees Fahrenheit.

(f) Now, go back to your Excel file and create a correlation matrix for these four variables, and place it underneath the SUMMARY OUTPUT.
(g) Re-save this file as: yield15
(h) Now, print out *just this correlation matrix* on a separate sheet of paper.

Answer to the following questions using your Excel printout. (Be sure to include the plus or minus sign for each correlation):

8. What is the correlation between RAINFALL and YIELD?
9. What is the correlation between NITROGEN and YIELD?
10. What is the correlation between TEMPERATURE and YIELD?
11. What is the correlation between NITROGEN and RAINFALL?
12. What is the correlation between TEMPERATURE and RAINFALL?
13. What is the correlation between TEMPERATURE and NITROGEN?
14. Discuss which of the three predictors is the best predictor of corn yield.
15. Explain in words how much better the three predictor variables combined predict corn yield than the best single predictor by itself.

Chapter 8: Practice Test

Let's consider an experiment in which houseflies were reared in separate culture jars in which the "medium" in each of the culture jars differed based on: (1) more water was added, (2) more sugar was added, and (3) more solid matter was added. You want to study the effect of the different media on the wing length of houseflies, measured in millimeters (mm). The hypothetical data are given in Fig. B.7.

WING LENGTH (mm) OF HOUSEFLIES IN DIFFERENT MEDIA

MORE WATER	MORE SUGAR	MORE SOLID MATTER
4.0	3.8	3.5
4.2	3.9	3.7
4.4	4.1	3.8
4.8	4.2	3.9
4.9	4.4	4.1
5.1	4.6	4.2
5.3	4.8	4.3
5.2	4.9	4.5
5.4	5.0	4.4
5.5		3.7
		3.9

Fig. B.7 Worksheet Data for Chap. 8 Practice Test (Practical Example)

(a) Enter these data on an Excel spreadsheet.
(b) On your spreadsheet, write the null hypothesis and the research hypothesis for these data
(c) Perform a *one-way ANOVA test* on these data, and show the resulting ANOVA table underneath the input data for the three types of media.
(d) If the F-value in the ANOVA table is significant, create an Excel formula to compute the ANOVA t-test comparing the more water added medium versus the more solid matter added medium, and show the results below the ANOVA

table on the spreadsheet (put the standard error and the ANOVA t-test value on separate lines of your spreadsheet, and use two decimal places for each value)

(e) Print out the resulting spreadsheet so that all of the information fits onto one page

(f) On your printout, label by hand the MS (between groups) and the MS (within groups)

(g) Circle and label the value for F on your printout for the ANOVA of the input data

(h) Label by hand on the printout the mean for more water added medium and the mean for more solid matter added medium that were produced by your ANOVA formulas

(i) Save the spreadsheet as: wing10

On a separate sheet of paper, now do the following by hand:

(j) find the critical value of F in the ANOVA Single Factor results table

(k) write a summary of the *result* of the ANOVA test for the input data

(l) write a summary of the *conclusion* of the ANOVA test in plain English for the input data

(m) write the null hypothesis and the research hypothesis comparing the more water added medium versus the more solid matter added medium.

(n) compute the degrees of freedom for the *ANOVA t-test* by hand for three types of media.

(o) use Excel to compute the standard error (s.e.) of the ANOVA t-test

(p) Use Excel to compute the ANOVA t-test value

(q) write the *critical value of t* for the ANOVA t-test using the table in Appendix E.

(r) write a summary of the *result* of the ANOVA t-test

(s) write a summary of the *conclusion* of the ANOVA t-test in plain English

Appendix C: Answers to Practice Test

Practice Test Answer: Chap. 1 (see. Fig. C.1)

WIDTH (at the widest part) OF HUMMINGBIRD EGGS

WIDTH (mm)		
9.4		
9.6	n	16
9.7		
9.8		
10.1	Mean	9.73
9.5		
9.4		
10.2	STDEV	0.30
9.6		
10.3		
9.6	s.e.	0.07
9.8		
10.1		
9.4		
9.5		
9.7		

Fig. C.1 Practice Test Answer to Chap. 1 Problem

Practice Test Answer: Chap. 2 (see. Fig. C.2)

FRAME NUMBERS	Duplicate frame numbers	RAND NO.
1	6	0.874
2	41	0.253
3	38	0.044
4	28	0.842
5	20	0.920
6	36	0.231
7	4	0.863
8	13	0.853
9	16	0.123
10	9	0.039
11	1	0.782
12	39	0.564
13	8	0.796
14	27	0.623
15	33	0.221
16	34	0.947
17	2	0.596
18	35	0.432
19	30	0.828
20	42	0.930
21	25	0.348
22	17	0.952
23	24	0.375
24	5	0.541
25	14	0.459
26	7	0.659
27	10	0.617
28	21	0.453
29	32	0.127
30	26	0.582
31	22	0.253
32	31	0.898
33	40	0.941
34	19	0.626
35	15	0.307
36	29	0.720
37	37	0.015
38	3	0.686
39	11	0.180
40	18	0.722
41	23	0.432
42	12	0.182

Fig. C.2 Practice Test Answer to Chap. 2 Problem

Practice Test Answer: Chap. 3 (see. Fig. C.3)

CONCENTRATION OF SO_2 IN THE ATMOSPHERE IN PARTS PER BILLION (ppb)								
ppb								
390		Null hypothesis:		μ	=	120 ppb		
332								
186		Research hypothesis:		μ	≠	120 ppb		
85								
29								
135		n		23				
86								
54		mean		73.13				
28								
35		stdev		100.25				
37								
28		s.e.		20.90				
18								
32								
24		95% confidence interval						
19								
21			lower limit		29.78			
35								
31			upper limit		116.48			
18								
20		29.78	----------------	73.13------	------------	116.48	----------------	120 -------
21		lower		mean		upper		Ref.
18		limit				limit		Value
Result:	Since the reference value is outside the confidence interval, we reject the null hypothesis and accept the research hypothesis.							
Conclusion:	The concentration of SO_2 in the atmosphere in the selected sites was significantly less this year than it was three years ago, and it is now probably closer to 73 parts per billion.							

Fig. C.3 Practice Test Answer to Chap. 3 Problem

Practice Test Answer: Chap. 4 (see. Fig. C.4)

BEAK DEPTH OF FINCHES

BEAK DEPTH (mm)					
6.1					
6.8	Null hypothesis:		μ	=	9.2 mm
10.3					
10.8	Research hypothesis:		μ	≠	9.2 mm
7.5					
9.8					
9.6	n	29			
7.2					
7.8					
9.9	MEAN	9.959			
7.6					
9.7					
11.4	STDEV	1.631			
8.3					
11.3					
11.4	s.e.	0.303			
11.3					
10.8					
10.3	critical t	2.048			
10.6					
10.8					
10.9	t-test	2.505			
11.1					
11.3					
11.2	Result:	Since the absolute value of 2.505 is greater than the critical t of 2.048, we reject the null hypothesis and accept the research hypothesis.			
11.4					
11.2					
11.3					
11.1	Conclusion:	The average beak length of survivor finches after a drought of six years was significantly longer than it was before the drought.			

Fig. C.4 Practice Test Answer to Chap. 4 Problem

Practice Test Answer: Chap. 5 (see. Fig. C.5)

NUMBER OF DAYS SPENT IN THE NEST

NORTH REGION	SOUTH REGION
95	90
98	92
99	94
112	93
103	95
105	98
106	100
109	102
110	103
111	104
112	98
114	102
95	99
98	
103	

Group	n	Mean	STDEV
1 North	15	104.67	6.53
2 South	13	97.69	4.53

Null hypothesis: $\mu_1 = \mu_2$

Research hypothesis: $\mu_1 \neq \mu_2$

1/n1 + 1/n2	0.14
(n1 - 1) x S1 squared	597.33
(n2 - 1) x S2 squared	246.77
n1 + n2 - 2	26
s.e.	2.16
critical t	2.056
t-test	3.23

Result: Since the absolute value of 3.23 is greater than the critical t of 2.056, we reject the null hypothesis and accept the research hypothesis.

Conclusion: Hedgehogs hibernated for significantly more days in the North Region than in the South Region (105 days vs. 98 days).

Fig. C.5 Practice Test Answer to Chap. 5 Problem

Practice Test Answer: Chap. 6 (see. Fig. C.6)

SITE	ELEVATION (m)	HEIGHT (cm)
1	1200	75
2	1350	65
3	1400	68
4	1550	60
5	1600	45
6	1750	50
7	1850	46
8	1900	48
9	2000	45
10	2100	25
11	2250	23
12	2300	20
13	2400	21
14	2850	18
15	3000	15

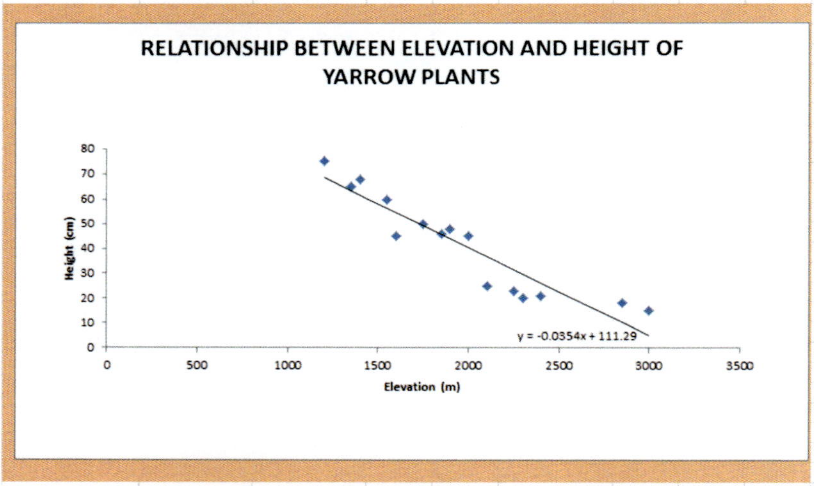

SUMMARY OUTPUT

Regression Statistics	
Multiple R	0.934
R Square	0.87
Adjusted R Square	0.86
Standard Error	7.45
Observations	15

ANOVA

	df	*SS*	*MS*	*F*	*Significance F*
Regression	1	4908.04	4908.04	88.43	3.65521E-07
Residual	13	721.56	55.50		
Total	14	5629.60			

	Coefficients	*Standard Error*	*t Stat*	*P-value*	*Lower 95%*
Intercept	111.293	7.66	14.53	0.00	94.75
X Variable 1	-0.035	0.00	-9.40	0.00	-0.04

Fig. C.6 Practice Test Answer to Chap. 6 Problem

Practice Test Answer: Chap. 6: (continued)

(d) a = y-intercept = 111.293
    b = slope = −0.035(note the negative sign!)
(e) Y = a + b X
    Y = 111.293 − 0.035X
(f) r = correlation = −.93 (note the negative sign!)
(g) Y = 111.293 − 0.035(2000)
    Y = 111.293 − 70
    Y = 41.29 cm
(h) About 23 − 25 cm.

Practice Test Answer: Chap. 7 (see. Fig. C.7)

CORN YIELDS ON IOWA FARMS

YIELD (bu/acre)	RAINFALL (in/yr)	NITROGEN (lbs/acre)	TEMPERATURE (degrees F)
240	28.8	190	80
255	29.5	195	79
299	27.6	230	79
230	28.3	175	78
300	31.5	250	79
180	27.1	250	82
294	28.9	220	78
278	27.1	200	74
284	28.8	204	80
194	30.7	185	81
170	27.9	190	88

SUMMARY OUTPUT

Regression Statistics	
Multiple R	0.78
R Square	0.602
Adjusted R Square	0.431
Standard Error	36.701
Observations	11

ANOVA

	df	*SS*	*MS*	*F*
Regression	3	14247.810	4749.270	3.526
Residual	7	9428.735	1346.962	
Total	10	23676.545		

	Coefficients	*Standard Error*	*t Stat*	*P-value*
Intercept	744.95	375.630	1.983	0.088
RAINFALL (in/yr)	6.33	8.283	0.764	0.470
NITROGEN (lbs/acre)	0.51	0.450	1.137	0.293
TEMPERATURE (degrees F)	-9.84	3.417	-2.880	0.024

	YIELD (bu/acre)	*RAINFALL (in/yr)*	*NITROGEN (lbs/acre)*	*TEMPERATURE (degrees F)*
YIELD (bu/acre)	1			
RAINFALL (in/yr)	0.19	1		
NITROGEN (lbs/acre)	0.31	0.03	1	
TEMPERATURE (degrees F)	-0.70	0.00	-0.05	1

Fig. C.7 Practice Test Answer to Chap. 7 Problem

Practice Test Answer: Chap. 7 (continued)

1. $R_{xy} = .78$
2. a = y-intercept = 744.95
3. $b_1 = 6.33$
4. $b_2 = 0.51$
5. $b_3 = -9.84$
6. $Y = a + b_1X_1 + b_2X_2 + b_3X_3$
   $Y = 744.95 + 6.33\ X_1 + 0.51\ X_2 - 9.84X_3$
7. $Y = 744.95 + 6.33(28) + 0.51(205) - 9.84(83)$
   $Y = 744.95 + 177.24 + 104.55 - 816.72$
   $Y = 1026.74 - 816.72$
   $Y = 210\,\text{bu/acre}$
8. +.19
9. +.31
10. −.70
11. +.03
12. +.00
13. −.05
14. The best predictor of corn yield was TEMPERATURE with a correlation of −.70. (*Note: Remember to ignore the negative sign and just use .70.)*
15. The three predictors combined predict corn yield much better ($R_{xy} = .78$) than the best single predictor by itself

Practice Test Answer: Chap. 8 (see. Fig. C.8)

WING LENGTH (mm) OF HOUSEFLIES IN DIFFERENT MEDIA

MORE WATER	MORE SUGAR	MORE SOLID MATTER
4.0	3.8	3.5
4.2	3.9	3.7
4.4	4.1	3.8
4.8	4.2	3.9
4.9	4.4	4.1
5.1	4.6	4.2
5.3	4.8	4.3
5.2	4.9	4.5
5.4	5.0	4.4
5.5		3.7
		3.9

Anova: Single Factor

SUMMARY

Groups	*Count*	*Sum*	*Average*	*Variance*
MORE WATER	10	48.8	4.88	0.27
MORE SUGAR	9	39.7	4.41	0.19
MORE SOLID MATTER	11	44	4.00	0.10

ANOVA

Source of Variation	*SS*	*df*	*MS*	*F*	*P-value*	*F crit*
Between Groups	4.06	2	2.03	10.86	0.0003	3.35
Within Groups	5.04	27	0.19			
Total	9.10	29				

More Water vs. More Solid Matter	
1/n1 + 1/n3	0.19
s.e.	0.19
ANOVA t	4.66

Fig. C.8 Practice Test Answer to Chap. 8 Problem

Practice Test Answer: Chap. 8 (continued)

Let MORE WATER = Group A, MORE SUGAR = Group B, and MORE SOLID MATTER = Group C.

(b) $H_0 : \mu_A = \mu_B = \mu_C$
$H_1 : \mu_A \neq \mu_B \neq \mu_C$
(f) $MS_b = 2.03$ and $MS_w = 0.19$
(g) F = 10.86
(h) Mean of MORE WATER = 4.88 and Mean of MORE SOLID MATTER = 4.00
(j) critical F = 3.35
(k) Result: Since 10.86 is greater than 3.35, we reject the null hypothesis and accept the research hypothesis
(l) Conclusion: There was a significant difference in the wing length of houseflies between the three types of media
(m) $H_0 : \mu_A = \mu_C$
$H_1 : \mu_A \neq \mu_C$
(n) $df = n_{TOTAL} - k = 30 - 3 = 27$
(o) $1/10 + 1/11 = 0.19$
s.e = SQRT(0.19*0.19)
s.e. = SQRT(0.036)
s.e. = 0.19
(p) ANOVA t = (4.88 − 4.00)/0.19 = 4.66
(q) critical t = 2.052
(r) Result: Since the absolute value of 4.66 is greater than the critical t of 2.052, we reject the null hypothesis and accept the research hypothesis
(s) Conclusion: The wing length of houseflies was significantly longer in the MORE WATER ADDED medium than the MORE SOLID MATTER ADDED medium (4.88 mm vs. 4.00 mm)

Appendix D: Statistical Formulas

Mean $\overline{X} = \frac{\sum X}{n}$

Standard Deviation $\text{STDEV} = S = \sqrt{\frac{\sum (X - \overline{X})^2}{n-1}}$

Standard error of the mean $\text{s.e.} = S_{\overline{X}} = \frac{S}{\sqrt{n}}$

Confidence interval about the mean $\overline{X} \pm t\, S_{\overline{X}}$

where $S_{\overline{X}} = \frac{S}{\sqrt{n}}$

One-group t-test $t = \frac{\overline{X} - \mu}{S_{\overline{X}}}$

where $S_{\overline{X}} = \frac{S}{\sqrt{n}}$

Two-group t-test

(a) when both groups have a sample size greater than 30

$$t = \frac{\overline{X}_1 - \overline{X}_2}{S_{\overline{X}_1 - \overline{X}_2}}$$

$$\text{where } S_{\overline{X}_1 - \overline{X}_2} = \sqrt{\frac{S_1{}^2}{n_1} + \frac{S_2{}^2}{n_2}}$$

$$\text{and where } df = n_1 + n_2 - 2$$

(b) when one or both groups have a sample size less than 30

$$t = \frac{\overline{X}_1 - \overline{X}_2}{S_{\overline{X}_1 - \overline{X}_2}}$$

$$\text{where } S_{\overline{X}_1 - \overline{X}_2} = \sqrt{\frac{(n_1 - 1)S_1{}^2 + (n_2 - 1)S_2{}^2}{n_1 + n_2 - 2} \left(\frac{1}{n_1} + \frac{1}{n_2}\right)}$$

$$\text{and where } df = n_1 + n_2 - 2$$

Correlation $r = \frac{\frac{1}{n-1} \sum (X - \overline{X})(Y - \overline{Y})}{S_x\, S_y}$

where S_x = standard deviation of X

and where S_y = standard deviation of Y

Simple linear regression

$Y = a + bX$
where a = y-intercept and
b = slope of the line

Multiple regression equation

$Y = a + b_1X_1 + b_2X_2 + b_3X_3 + \text{etc.}$
where a = y-intercept

One-way ANOVA F-test

$F = MS_b/MS_w$

ANOVA t-test

$$ANOVA\ \ t = \frac{\overline{X}_1 - \overline{X}_2}{s.e._{ANOVA}}$$

where $s.e._{ANOVA} = \sqrt{MS_w\left(\frac{1}{n_1} + \frac{1}{n_2}\right)}$

and where $df = n_{TOTAL} - k$
where $n_{TOTAL} = n_1 + n_2 + n_3 + \text{etc.}$
and where k = the number of groups

Appendix E: t-Table

Critical t-values needed for rejection of the null hypothesis (see Fig. E.1)

Fig. E.1 Critical t-values Needed for Rejection of the Null Hypothesis

sample size n	degrees of freedom df	critical t
10	9	2.262
11	10	2.228
12	11	2.201
13	12	2.179
14	13	2.160
15	14	2.145
16	15	2.131
17	16	2.120
18	17	2.110
19	18	2.101
20	19	2.093
21	20	2.086
22	21	2.080
23	22	2.074
24	23	2.069
25	24	2.064
26	25	2.060
27	26	2.056
28	27	2.052
29	28	2.048
30	29	2.045
31	30	2.042
32	31	2.040
33	32	2.037
34	33	2.035
35	34	2.032
36	35	2.030
37	36	2.028
38	37	2.026
39	38	2.024
40	39	2.023
infinity	infinity	1.960

Index

A
Absolute value of a number, 66–67
Analysis of Variance
  ANOVA t-test formula (8.2), 177–184
  degrees of freedom, 178–179
  Excel commands, 179–182
  formula (8.1), 175, 178
  interpreting the Summary Table, 175, 179
  s.e. formula for ANOVA t-test (8.3), 192, 244
ANOVA. *See* Analysis of Variance
ANOVA t-test. *See* Analysis of Variance
Average function. *See* Mean

C
Centering information within cells, 6–8
Chart
  adding the regression equation, 142–144, 148, 152, 229
  changing the width and height, 5–6, 155
  creating a chart, 121–131
  drawing the regression line onto the chart, 121–131
  moving the chart, 129–130
  printing the spreadsheet, 131–133
  reducing the scale, 132
  scatter chart, 123
  titles, 123–125, 127
Column width (changing), 5–6
Confidence interval about the mean
  drawing a picture, 43
  formula (3.2), 39–40
  lower limit, 36–37, 40, 44, 63
  upper limit, 36–37, 40, 44, 63
  95% confident, 36–37, 39–46, 76, 225
CORREL function. *See* Correlation
Correlation
  formula (6.1), 114, 144
  negative correlation, 109, 111, 112, 140, 145, 151
  positive correlation, 109, 110, 111, 116, 120, 145, 150, 151, 163
  9 steps for computing, 114–116
COUNT function, 9, 52
Critical t-value, 59, 72, 102, 105, 107, 179, 184, 186, 188, 246

D
Data Analysis ToolPak, 134–137, 154, 171
Data/Sort commands, 26
Degrees of freedom, 85, 86, 88, 89, 91, 100, 178–179, 184, 186, 188, 220, 222, 232

F
Fill/Series/Columns commands
  step value/stop value commands, 5, 22
Formatting numbers
  currency format, 15–17
  decimal format, 138

H
Home/Fill/Series commands, 4, 22
Hypothesis testing
  decision rule, 52, 66–67, 85, 176–177, 179

T.J. Quirk et al., *Excel 2016 for Biological and Life Sciences Statistics*, Excel for Statistics, DOI 10.1007/978-3-319-39489-3

Hypothesis testing (*cont.*)
- null hypothesis, 48–61, 63, 66, 69, 70, 74, 77–79, 84, 86–90, 91, 93, 96, 98, 99, 104, 105, 106, 107, 176, 178, 179, 183, 185, 186, 187, 188, 218, 220, 222, 225, 226, 227, 231, 232, 243, 246
- rating scale hypotheses, 48–51, 70
- research hypothesis, 48–54, 56–58, 60, 61, 63, 66, 69, 70, 74, 77, 79, 84–91, 93, 96, 98, 99, 103, 104, 105, 107, 176, 178, 179, 183, 185, 186, 187, 188, 218, 220, 222, 225–227, 231, 232, 243
- stating the conclusion, 53, 54, 57
- stating the result, 57
- 7 steps for hypothesis testing, 51–57

M

Mean
- formula (1.1), 1, 3, 37, 39

Multiple correlation
- correlation matrix, 160–164, 165, 167, 169, 230
- Excel commands, 4, 27, 179–182

Multiple regression
- correlation matrix, 160–164, 165, 167, 169, 230
- equation (7.1), (7.2), 153–160, 165, 167, 169, 230, 245
- Excel commands, 4, 27
- predicting Y, 153

N

Naming a range of cells, 8–9

Null hypothesis. *See* Hypothesis testing

O

One-group t-test for the mean
- absolute value of a number, 66–67
- formula (4.1), 65, 67, 91, 144
- hypothesis testing, 65–69, 76
- s.e. formula (4.2), 91
- 7 steps for hypothesis testing, 65–69

P

Page Layout/Scale to Fit commands, 30, 44

Population mean, 35–36, 38, 48, 4965, 67, 84, 91, 171, 176, 177, 178, 179, 181

Printing a spreadsheet
- entire worksheet, 46, 74, 96, 145–147
- part of the worksheet, 145–147, 223
- printing a worksheet to fit onto one page, 131

R

Random number generator
- duplicate frame numbers, 24–28, 33, 34, 224
- frame numbers, 21–24, 33, 224
- sorting duplicate frame numbers, 26–29, 34

RAND(). *See* random number generator

Regression, 4, 27, 109–165, 167, 169, 230, 245

Regression equation, 245
- adding it to the chart, 142–144, 148, 152, 229
- formula (6.3), 245
- negative correlation, 109, 111, 112, 140, 145, 151
- predicting Y from x, 153
- slope, *b*, 140
- y-intercept, *a*, 140
- writing the regression equation using the Summary Output, 137–140, 145, 147, 158, 159

Regression line, 121–131, 140–145, 148, 150, 151, 152, 229

Research hypothesis. *See* Hypothesis testing

S

Sample size
- COUNT function, 9, 52

Saving a spreadsheet, 13

Scale to Fit commands, 30, 44

s.e. *See* Standard error of the mean

Standard deviation
- formula (1.2), 2, 244

Standard error of the mean
- formula (1.3), 3, 10, 244

STDEV. *See* Standard deviation

T

t-table. *See* Appendix E

Two-group t-test
- basic table, 83
- degrees of freedom, 86
- drawing a picture of the means, 89
- formula (5.2), 91
- Formula #1 (5.3), 92–97
- Formula #2 (5.5), 98–102
- hypothesis testing, 82–90
- 9 steps in hypothesis testing, 82–90
- s.e. formula (5.3), (5.5), 3, 10, 244